A Daughter of Witches

A Romance

Joanna E. Wood

Writat

This edition published in 2022

ISBN : 9789356395336

Contents

CHAPTER I

Miss Temperance Tribbey stood at the back door of the old Lansing house, shading her eyes with one hand as she looked towards the gate to discover why Grip, the chained-up mastiff, was barking so viciously.

The great wooden spoon, which she held in her other hand, was dripping with red syrup, and showed that Temperance was preserving fruit. To the eyes of the initiated there were other signs of her occupation. Notably a dangerous expression in her eyes. The warmth of the stove was apt to extend to Miss Tribbey's temper at such times.

Sidney Martin, coming up the avenue-like lane to the farm-house, did not observe Miss Tribbey standing at the back door, although she saw him; and, therefore, much to his own future detriment and present prejudice in Miss Tribbey's eyes, he went to the front door, under its heavy pillared porch, and knocked. After he had vanished round the corner of the house towards the ill-chosen door, Miss Tribbey waited impatiently for the knock, calculating whether she could safely leave her fruit on the fire whilst she answered it.

The knock did not come. Muffled by the heavy door, its feeble echo was absorbed by the big rooms between the front door and the kitchen.

"Well!" said Temperance, "has he gone to Heaven all alive, like fish goes to market, or is he got a stroke?"

The cat arched its back against Miss Tribbey's skirts and so shirked the reply which clearly devolved upon it, there being no other living creature visible in the big kitchen.

"It's as bad as consumpting to have a man hanging over a body's head like this," continued Temperance. "My palpitations is coming on! If I'm took with them and that fruit on the fire, along of a man not knowing enough to knock!"

The fruit in the big copper kettle began to rise insidiously towards the brim.

"I'll just go and take a speck at him through the shutters," said she.

She crossed the kitchen, but ere she left it, long housewifely habit made her "give a look to the stove." The burnished copper kettle was domed by a great crimson bubble, raised sphere-like by the steam.

"My soul!" said Temperance, and took a flight across the kitchen, lifting the heavy pot with one sweep from the fire to the floor. The dome quivered, rose a fraction and collapsed in a mass of rosy foam.

The crisis was past, and just then the expected knock came.

Temperance drew a long breath.

"There!" she said, "that jell's done for! I'll have to stand palavering with some agent chap or book-canvasser with my jell a-setting there gettin' all muddied up."

This reflection bore her company to the front door, which she opened with an air of calm surprise. Miss Tribbey knew her manners.

"Well, I declare!" she said. "Have you been here long?"

"No—came this very minute," said Sidney in his soft, penetrating voice.

"Oh, the liar!" said Miss Tribbey to herself, scandalized.

"It's beautiful here," he continued. "That field of yellow grain there is worth a journey to see."

("Poor crittur," Miss Tribbey said in relating this afterwards. "Poor, ignorant crittur! Not knowing it's a burning, heartsick shame to see grain that premature ripe with the hay standing in win'rows in the field, before his eyes.")

"Ahem!" said Miss Tribbey, her visitor showing signs of relapsing again into that reverie which had made the interval of waiting seem as nothing to him, unconscious as he was of the narrowly averted tragedy with Miss Tribbey's fruit.

But face to face with her he was too sensitive not to recognize her impatience with his dallying mood. He roused himself and turned towards her with a frank and boyish smile.

"I'm bothering you," he said, "and doubtless keeping you from something important."

"I'm making jell," said she briefly, her attitude growing tense.

"Have you heard Mr. Lansing speak of Sidney Martin?" he asked. "In reference to his coming to stay here this summer? I'm Sidney Martin, and I want to come, if it is convenient to receive me, the beginning of next week, and——"

"Come where?" demanded Temperance.

"Here," said Sidney, a little embarrassed.

"To this house?"

"Yes," said Sidney, looking at her with the confidence in his eyes of one who, loving his fellow-creatures more than life, expects and anticipates their love in return; one does not often see this expression, but one often sees the residuum left after the ignorance it bespeaks has been melled and mingled by experience.

"Mr. Lansing is over at the unction sale at Abiron Ranger's," said Temperance. "You'll excuse me, my jell's a-waitin' for me, and whatever time other folks has to waste I've none! You'll excuse me! I know nothing about boarders and sich!"

"Boarders," said Sidney in alarm, looking about for signs of the enemy. "Do you take boarders?"

"It would seem so now," said Temperance, cuttingly. "It would sertingly seem so."

"Oh, bless you!" said Sidney. "I'm not a boarder! I'm a visitor. There's a great difference, isn't there? I'm the son of old Sidney Martin, the county clerk who went away to Boston and married there. You have heard of him, haven't you?"

"Yes, I have," said Temperance, throwing one end of her apron over her head to shield off the sun. "Yes, I have, though I was in short petticoats and my hair in a pig-tail when it happened. He went to Bosting and married rich, didn't he?"

"He married in Boston," said Sidney. "Where is Mr. Ranger's?"

"Abiron Ranger lives two miles down the road, across to the right," said Temperance. "He died a week ago Wednesday, and there's an unction sale there to-day. There's goin' to be a divide up. The widow wants her thirds. A very pushin' woman Mis' Ranger is."

"Two miles more," said Sidney, with something like a sigh.

Miss Tribbey's keen eyes noted that he was white as from recent illness.

"Won't you set down a spell and hev' a glass of milk?" she asked; "set down in the shade there, and I'll get it in a jiffy. What's the sense of standing in a blazing sun like this?"

She whisked off and presently returned with the milk and a plate of New England cookies.

"I've got to go back to my jell," she said. "When you get ready to go just put the things on the porch. My soul! I was took when you began talking about boarders. For I've said, and said often, 'If boarders comes, I go.' But visitors! We've always heaps of company, and I'll go bail no house I do for'll ever be took short of things to put on the table; the most unexpectedest company that ever drove up that avenue was always set down to a liberal table; when you go down the road about a mile, just look towards the right, and you'll see a brown frame-house, with a lightning-rod on it. That's Abiron Ranger's. Cut across the fields. It's shorter."

"Thanks," said Sidney; "what delicious milk."

"Yes—Boss don't give chalk and water, she don't," said Miss Tribbey, and went off to her kitchen.

"A poor, slim jack of a man," she soliloquised, ladling out her jelly. "My soul! There's a mighty difference between him and Lanty—but there—his kind don't grow on every bush. Clear Lansing he is, through and through, and there never was no runts among the Lansings."

For a few minutes Sidney rested in the porch, his eyes dwelling upon the undulating grain before him. To one more experienced in these matters, its burnished gold would have told sad tales of the terrible drouth which had scorched the country side, but to him it appeared the very emblem of peace and plenty.

What field of the cloth of gold was ever equal in splendour to this?

He rose and passed down the dusty road. Upon one hand the panicles of an oat field whispered together, upon the other stretched the barren distances of a field known far and near as Mullein Meadow, these weeds and hard shiny goat grass being all that grew upon it.

Sidney did not grasp the significance of its picturesque grey boulders, nor think how dear a possession it was at the price of the taxes upon it. After Mullein Meadow came a little wood, thick with underbrush, in the shadow of which a few brackens were yet green; and fronting the wood a hayfield, with a patch of buckwheat in full bloom in one corner, showing against the dim greenness of the hay like a fragrant white handkerchief fallen from an angel's hand.

Sidney cut across the hayfield to where a glistening point glimmered in the sunshine, above a sloping roof set on brown walls.

"'How curious. How real!'" he said to himself. "'Underfoot the divine soil—Overhead the sun.'"

He reached the enclosure in which the house stood, and paused at the gate to watch the groups of men discussing their purchases, for the sale was over.

Presently, his interest urging him, he entered, and mingled with the others, having the fanciful idea he would know his father's old friend by intuition. His eyes softened as he looked at the weather-beaten faces and hard-wrought hands of these men. The memory of the golden grain was dimmed a little, and he saw bands of men bending above their toil beneath stern skies, "storing yearly little dues of wheat, and wine, and oil." But that vision was illy entertained in his sanguine, idealistic imagination. It was dismissed to give place once more to the "free farmer" of song and story, and as if to bear witness to this latter picture, a young man detached himself somewhat impatiently from a group of his fellows, and advancing towards where Sidney stood, flung himself across a mettlesome roan which was tugging viciously at its bridle as it stood tied beneath a tree.

The young man's face was flushed, he was blue-eyed and debonair, his yellow hair tossed back carelessly above his brow; a wide, flapping felt hat rested on the back of his head. His features were large and strongly carved. His mouth, seen red through his tow-coloured moustache, had all the sweetness of a woman's and much of the deviltry of a rake's. But his face did not look vicious, only dangerous. His strong, lean hand curbed his horse easily; he turned in his saddle to call to those whom he had just left.

"If anyone wants a last word he knows where to find me," he cried.

"Yes," someone said, giving a coarse laugh; "near some pair of apron strings."

"What did you say?" demanded Lanty Lansing, urging his horse near the group.

"Nothing; O, nothing, Lanty," said the speaker irritatingly, whilst Lanty's horse circled the group crab-fashion.

"Don't let me keep you," went on the man, and Sidney saw he was heavy, black-browed and strongly built. "Don't let me keep you. Is it the little yellow-haired one or the other? I like the little one best myself, Miss——"

"Keep my cousins' names out of your mouth," said Lanty, his quick temper in a flame, "or I'll break your neck for you."

"If all's true that's told for true, you're better at breaking hearts than necks. There's a little girl over Newton way—but there! I'll tell no tales; but to say you're going to have both your cousins! You're a Mormon, Lanty, that's what you are."

"Be quiet! Be quiet!" some of the men said. "Good-bye, Lanty, better be off; he don't mean nothing."

But the big man, sure of the prestige of his size, thinking, evidently, that Lanty would not dismount, was not to be silenced. Perhaps he was hardly quite sober. He was a machine agent in the neighbourhood, and had bidden unsuccessfully against Lanty for a horse. His next words took him too far.

"I ain't sayin' anything to put his back up! All I say is that them cousins of his can smile at other folks as well as at him, and why shouldn't they? I don't like a girl no less because she——" He never finished.

Lanty was off his horse like a flash. His fist caught the big man under the jaw, lifting him almost off his feet and sending him crashing down. Lanty waited with hands clenched for him to rise. The crowd swayed, those close at hand giving way, those upon the outskirts pressing forward. The horse, so suddenly released, reared and swung round on its hind legs, and just then Sidney saw a tall, finely-formed young woman appear almost under the plunging horse, twist a strong hand in the bridle, and wring both curb and snaffle so viciously that the beast gave his head to her guidance. She wheeled it towards Lanty.

"Lanty," she said, laying her free hand upon his arm; "Lanty."

"Go into the house, Vashti," he said. "What are you doing here?"

"You are not going to fight," she said, "*with him*?"

Lansing was silent; she continued: "Go home, Lanty, please——"

Some of the older men had closed around the man, who was just rising to his feet. The first mad impetus of battle was cooling in Lansing's veins, and just then another girl pushed through the group of men to his side. A slight, graceful creature, with the Lansing blue eyes and fair skin, and sweet lips, she was trembling—white.

"Lanty," she said, with terror in her eyes, "has he hurt you? You frighten me horribly."

His eyes rested upon her, self-reproach making them eloquent.

He took the reins from the tall girl's hand, looking always at the white, appealing face of the other.

"I'm a bit of a fool," he said; "but he spoke of you two and——" he paused; "I'll be over to-night," he said, and rode away.

Vashti Lansing's hand and wrist were wrenched, and already beginning to swell; the rearing horse had not been so easily subdued after all; but physical pain was a slight thing to her just then.

"Come," she said to her cousin. "There's father coming out. Don't tell him; let some one else. There are always plenty ready to talk."

So the two girls went into the house, and Sidney Martin stood gazing after them, rapt in the vision of a magnificently made woman curbing and subduing a rearing horse. Surely a type of the eternal divinity of womanhood, striving against the evil that men do.

Sidney Martin, dreamer of dreams, cherisher of ideals, delicate and supra-sensitive, was subjugated for ever by love of that splendid piece of vitality—that woman whom even at this moment he likened to the *Venus de Milo*—whose magnificent energy and forceful grace recalled so vividly the *Winged Victory of Samothrace.* With a throb he remembered the beauty of that headless masterpiece, where it stands in the cool greyness of the Louvre, the inexplicable sense of triumphant effort expressed in its heroic pose. How many aspiring souls have gathered fresh courage from its mutilated majesty, where it stands at the head of the wide stairs! And here in the New England hills he had found a woman who might have been its original. The great sculptors had not dreamed then, when they created these Goddesses of stone; even unto this day, it seemed, they walked the earth, radiant in strength and beauty. How fitting that their statues should be given wings, to typify the splendid spirit prisoned in the imperial clay!

Sidney watched the girls enter the house and followed them involuntarily. As he passed the bully who had been knocked down, he saw there was a lump like an egg already adorning his jaw.

"Serves him right!" said gentle Sidney Martin to himself.

A little farther on half-a-dozen men stood talking together; one, whom Sidney took to be the auctioneer, was saying meaningly:—

"It's a bad business bidding for what you don't want."

"Yes," said one of the group, laboriously keeping up the joke. "Yes—for it's apt to be knocked down to you, and then you've got it on your hands."

"On your jaw, you mean," said the auctioneer smartly; whereat a laugh went up. Clearly Lanty Lansing had partizans here.

As Sidney reached the doorway within which the girls had vanished, a grey-haired man stepped forth from it.

"Mr. Lansing?" said Sidney, confidently.

His intuitions had not played him false.

"I am Sidney Martin," he continued, but got no further with his self-introduction.

"I'm right glad to see you," said old Lansing heartily, "right glad to see you! So you're old Sid's son? Well, you don't favour him no more than my girl favours me! I was struck all of a heap when Dr. Clement told me he knew old Sid's son in Bosting; says I, 'If that younker is like his father I should say he'd have a liking for the fields, even if he is Bosting born and bred.' But there! How did you come? Is your things at the station? How long is it since old Sid died? A nice old boy was Sid! And he had a talent for finding wood-chucks that beat the dickens."

"I lost my father four years ago," said Sidney—"he often spoke of you."

"I'll warrant he did," said old Lansing, "and my girls know old Sid as well as their next door neighbour. Sid weren't one of the sort to go back on old times—girls"—raising his voice—"girls!"

The two girls reappeared side by side.

"This is Mr. Martin," said the old man; "Old Sid Martin's son."

The girls gave him characteristic salutations. Vashti's inclination was stately, with all the plastic grace of her beautiful form expressed in it. Mabella, to whose cheeks the soft rose had returned, bestowed upon him the tantalizing salutation of the born coquette, piquant, confident, but withal reserved.

"My daughter Vashti," continued Mr. Lansing. "My niece Mabella"—and then—"Where's Lanty?"

"He has gone home," said Vashti; her voice was soft and full; a rarity in that region, and a heritage from the Lansings of old.

"That's too bad! It's my nephew—Lansing Lansing," he went on to Sidney, "the last of the Lansings."

"He's coming over to-night," said Mabella.

"You'll see him then; there are only four Lansings left now. An old man, a young one, and two girls. Well, it's a good old stock and that's plenty for a fresh graft. Well, well! How's Miss' Ranger feeling, girls? Are you ready to go home?"

"Yes—quite. She's more cheerful," said Vashti. "Shall we get ready?"

"Yes." Then turning to Sidney: "Where did you say your things were?"

Sidney had no time for explanations up to this moment.

"I've been staying at Brixton," he said; "and this afternoon I thought I would come over and ask when I might come, as you had been so good as to invite me; so I came by train from Brixton, and walked to Lansing Farm, and there I saw a lady who directed me here."

"Temperance," said the two girls, and looked at each other.

"She gave me some delightful milk and her opinion of boarders," said Sidney, smiling.

Mabella's laugh rang out like the call of a bird.

"Go and get ready, girls," said Mr. Lansing, "and I'll fetch the horses around."

The girls went indoors, first telling Sidney they would not be long. He went to the side of an old well and sat down upon the edge, looking into its cool depths; far, far down, he could see the distorted vision of his own face. A fat toad hopped lazily about the stones in the moist coolness of the well mouth. The wooden handle of the windlass was worn by many palms—as the creeds of the world are fretted and attenuated by the very eagerness of those who seek the Living Waters by their aid. Hop-vines grew over the house and Phœbe birds fluttered through their rustling leaves. The men stared curiously at the stranger by the well, to whom presently came the two girls again, in flat, wide hats.

"How brave you are!" he said to Vashti, rising at their approach. He was more than ordinarily tall, but Vashti's stately head was well above his shoulder.

"How brave you are! That beast of a horse looked frightful as it stood rearing above you! I thought you would be killed."

"I am not afraid of many things," said Vashti, soberly. Yet there was that leaping within her breast which sometimes frightened her sorely.

Sidney's eyes dilated with eagerness as he drank in the suave beauty of her statuesque shape. It was a beauty which appealed to him keenly. Divorced from all minor attributes, it depended securely upon form and line alone; colour, environment, counted as nothing in its harmonious whole. But one of the flexile wrists was swollen and stiff.

"You are hurt," he cried, forgetting that to keen eyes his anxiety might seem absurd. "You are hurt! That horse!"

She coloured a little—slowly—it was like the reflection of a rosy cloud on marble.

"Yes, it is twisted, I think."

He looked at it and shuddered. It gave him a sense of absolute physical nausea to see suffering. He had had a strange bringing-up by a visionary mother, who, absorbed in a vision of the Pain of the world, had impressed her morbid ideas upon her child, until now, in manhood, he was as sensitive to even the abstract idea of pain as the eye is to dust. Before real suffering his whole being shrank. At that moment Mr. Lansing drove up in the democrat waggon; but a change which was very apparent had come over his countenance.

Vashti and Mabella looked at each other and nodded apprehensively.

"Get in, girls," said the old man in abrupt, authoritative tones. "Come up beside me here," he said to Sidney.

They drove through the yard in silence, old Lansing nodding good-bye curtly to his neighbours. The moment they were on the road he turned to the two girls:

"What's this I hear?" he demanded. "Lanty has been fighting again! Verily 'he that slayeth with the sword shall perish by the sword.' It's a scandal."

"It wasn't a sword; 'twas his fist," said Mabella *sotto voce.*

"He only knocked the man down," said Vashti, "and he needed it."

"You're a judge of such things, evidently," said her father irately. "I say it's a disgrace to be a common brawler—to——"

Mabella spoke up eagerly. "Oh, but uncle," she said. "The man said something about Vashti and me—I don't know what, but not pleasant, and——"

"He did, did he?" demanded the old man, his face growing strangely like Lanty's in its anger. "He *did.* Wait till I see him! I'll break every bone in his body if I catch him"; he cut the horses viciously with his whip. "Only wait." Evidently he had forgotten his doctrine of peace. As a sky is lighted by an after-glow into the beauty of dawn, so old Lansing's face illumined by his wrath was youthful once more.

"He spoke slightingly of you, did he! The——"

He choked down an unscriptural epithet.

Mabella nudged Vashti gleefully. Sidney managed to give the girls a look of sympathetic congratulation over his shoulder. But Vashti's face was still sombre. She knew her father far too well to think he would be consistently inconsistent. Lanty would have his bad half-hour, irrespective of this raging. The Lansings were essentially illogical. It was a common saying in the neighbourhood, that calculating upon a Lansing was like catching a flea: when you thought you had him he wasn't there!

"What did you think when Dr. Clement gave you my invitation?" asked old Lansing.

"I was simply delighted," said Sidney; "you know I did not feel that I would be coming among strangers. My dear old dad spoke very often of the Lansings, and you in particular."

"Yes—he wanted sister Mabella, her mother there; we quarrelled, sister and me, over that matter. She would have her cousin Reuben and nobody else. Poor things! Neither the one nor the other of 'em lived long. We Lansings are great on marrying cousins," he said half apologetically, suddenly remembering it was this young man's father who had been slighted.

Both the girls blushed, but the blush died unseen upon the cheeks of each. For neither searched the countenance of the other. How blindly we stumble towards our own desires—unheeding the others who seek the same treasure till a hand plucks it away from before us, and then with empty hands we brush the mists from our eyes, and see how, led by fatuous delusion through perilous places, we have come to the ashes! But the ashes are never so dead that eager breath may not blow them into that Phoenix flame whence Hope is born.

"My father told me all the old stories of his boyhood," said Sidney. "I have heard of all your adventures together."

"So have we," said Mabella. "Do you know the story of how a streamer of crape was tied upon the door of the old church the night the Independents opened theirs?"

"Yes, indeed," said Sidney, laughing. "My father related that exploit of Mr. Lansing's many a time."

Both girls laughed aloud, at least Mabella did, and Vashti's full, soft laughter echoed through it like the call of a wood dove.

"My uncle," said Mabella, with emphasis, "has told us how your father did it."

"Tut—tut," said old Lansing, not ill pleased. "Not worth repeating—school-boy capers."

Afterwards in comparing notes the girls and Sidney found that in every instance the teller of the story had given the other the hero's rôle to play. A generous thing, surely. Yet, like all true generosity, not barren. For in the imaginations of all these young people, this Damon and Pythias of the New England hills shared a dual glory for deeds of "derring do" against scholastic authority and ghostly reverence; and their names went down to posterity as mighty hunters of the woodchuck.

"Must you really go back to Brixton to-night?" asked Vashti of Sidney, as they alighted from the democrat waggon. The man trembled as he looked upon her, so strongly had her individuality impressed him.

"Yes," he said. "I must go back to-night, but," he added, not concealing his eagerness, "I shall return."

"Whenever you can, and the sooner the better," said old Lansing, interrupting him.

"Monday, then," said Sidney.

"Monday be it," replied the old man, pleased with his eagerness. "You want to get browned up a bit," he added. "Have you been ill?"

"Grippe—in the winter," Sidney Martin said, suddenly feeling ashamed of acknowledging it—before that splendid creature whose presence seemed such a reproach to all less superbly well than herself. It was a bad sign, had Sidney been looking for such subtleties, that Vashti's magnificence of physique impressed him as a reproach against imperfection, rather than as a triumph of the race. It was so with her always. She gave others a chilling sense of what the human "might have been" rather than an inspiring perception of what the human "might be." Surely the spirit is subtly strong, giving each individual an *aura* of his own which may stimulate those who enter it like the piney ozone of the mountains, or stifle them as does the miasmic breath of a morass.

"Well—if you must really go——" said Mabella.

Supper was over—a supper presided over by Temperance Tribbey, and justifying thoroughly her remarks upon her capability as a purveyor. Sidney was taking leave at the front door preparatory to his departure for the station.

"Yes—don't keep him any longer, girls. He'll miss his train. It is sun-down now; another dry sun-down at that! It's killing weather. Well, good bye—we'll look for you Monday."

"Yes, on Monday," said Mabella's treble.

"On Monday," echoed Vashti's contralto.

"On Monday," repeated Sidney, raising his hat and turning away, and the voices of the three blent even as their lives were to do.

At the gate Sidney turned; Mabella had vanished promptly to adorn herself against the arrival of Lanty. The old man had gone off to the stables.

Vashti stood alone, her figure erect beside the Corinthian pillar of the old colonial porch. The rigid line of the column accentuated the melting curves of shoulder and hip. Lighted by the yellow after-glow she seemed transfigured to his glamoured fancy. He bared his head, and the goddess raised her hand in farewell. He passed down the road in a dream, hardly noting Lanty, as he rode past him to where Vashti waited in the after-glow.

CHAPTER II

He was to see her daily during the summer, breathe the same air with her, commune with her familiarly, and in a measure share the same experiences. This had been all Sidney Martin's thought, from the time he left Vashti Lansing haloed by the yellow after-glow, until the Monday following, when he entered the avenue leading up to the Lansing house.

This time he and his belongings had been driven over from Brixton. The drive had been long—a good ten miles, over dusty mountain roads, between fields crisped and parched by the pitiless sun; but at every turn of the road Sidney's fanciful imagination had figured forth a radiant form which beckoned him on. How sweet the welcoming sign would be when the farewell gesture had been so gracious! And now he had arrived. When, where would he see her first? Would the glory of the setting sun have left her face? Would she—and then he saw her.

In the wide angle made by the wing of the house there grew a great mass of hollyhocks, perfumeless, passionless flowers, fit for the garden of Proserpine. They were in full bloom. Not the pin-cushiony, double flowers of the "improved"—save the mark!—hollyhock, but the exquisite, transparent, cup-like single ones. In every shade, from crimson to pink, from salmon to white, from lemon to a rich wine colour, they grew there, stiff, stately, severe, their greyish green foliage softening the brilliancy of their blossoms. Scores of yellowish-white butterflies fluttered about them, sometimes entering boldly to the heart of the flowers, sometimes poising upon the button-like buds which crowned the tapering stems. And in the midst of this pure sweetness stood Vashti.

Sidney sprang from the musty carriage and went towards her, going, as it seemed to him, into a more exalted atmosphere at a step.

And as he saw her then, he saw her ever afterwards—not, perhaps, wholly as man looks at woman, rather as the enthusiast views perfection, as the devotee adores the Faith made visible. He saw her not as an individual woman, but as the glorious typification of her sex.

Ah, mysterious medley of mind and body! Ah, pitiful delusion which suggests a sequence of spirit and shape!

She gave him her hand cordially enough, not a small hand, but one exquisitely proportioned to her stature.

"We are so glad to see you," she said. "Father is in the far-away meadow at the hay; Mabella will be here in a moment."

"Is your hand better?" he asked.

"Yes, oh quite!" she replied, pleased that he had remembered.

Temperance and the driver carried the trunk upstairs; the driver departed and Temperance came to greet Sidney. It was afternoon, and Temperance was busy at her patchwork. She sewed dexterously while she talked.

"Terrible weather, ain't it?" she began. "My soul! Seems to me the Lord's clean forgot us here. The paint on the shed's fair blistered, and the cat's thin with the heat. The grain's done for, and the hay's no better'n rakings, and as for the roots—well, there'll be none if it don't rain, and do it quick, too. 'Drink, and praise God,' the preacher's got painted on his well by the way, and the well's been dry these five weeks. Look at that sky! It's dry as bass-wood. My chickings is going about with their mouths open, and there's nothing in the ponds but weeds and frogs. They say frogs grow in water, but I never seen the beat of the frogs this year. They say the Frenchers eats 'em. It's a pity our men couldn't learn, and we'd pay a sight less for butchers' meat. My soul!"—Temperance's lecture upon the drought was brought to an abrupt conclusion. Mabella, not seeing Sidney standing in the shadow, had come stealing up behind Miss Tribbey, and suddenly seizing her round the waist swung her round in a breathless whirl.

"My soul!" said Miss Tribbey again, releasing herself violently, and feeling her head and patting her person mechanically, as if to be certain she was intact. "You ain't bridle-wise yet, M'bella. It's cur'us you don't seem to get sense."

Mabella laughed.

Miss Tribbey continued with an ill-sustained show of bad temper, "You kin laugh, but it's discouragin'."

"It is," agreed Mabella blithely. "I'm like Nathan Peck."

"Go 'long with you!" said Temperance, tossing her head. "Nathan ain't none too brainy, but I never seen any such carryin's on as them with him!"

Temperance beat a retreat to the kitchen. Mabella laughing turned to Vashti, and for the first time caught sight of Sidney.

"Oh!" she said with a little gasp; then pulling herself together, advanced with outstretched hand. The ready rose dyed her cheek. She looked like some pretty culprit child. Her eyes were blue as a gentian flower—"Lansing blue" the neighbours called them. Her mouth expressed all the sweetness of a pure and loving nature. Her air was full of blithe gaiety. She seemed the incarnation of summery youth. There was something in her manner, too, of tremulous excitement—as of one not yet knowing life's secret, but in sight of the mystery, eager yet afraid of passing its portal.

Sidney was greatly won by her pretty air of deprecation, which mutely apologized for the small whirlwind she had created by her entrance.

"Come," said Vashti to Sidney. "It's too bad to keep you standing here." So they left the hollyhocks.

"Who is Nathan Peck?" asked Sidney of Mabella.

"Temperance Tribbey's beau," she said with a little laugh. "They've been keeping company for nineteen years."

"Don't they know their own minds yet?"

"Nathan does, but Temperance doesn't believe in being hasty," said Vashti with what, in a less majestic creature, might have been a sneer.

"And to tell the truth she doesn't want to leave us," said Mabella, who invariably found the best motive for other people's actions. "She's the dearest old thing!"

"Father declares," said Vashti, "that you are to do exactly as you like. He's working at the hay. They're working late now and we take them out something to eat at four o'clock. If you would like to come with us——"

"Oh, yes," said Sidney, "I should like it of all things."

"Well, we'll be going in half an hour or so. But wouldn't you like to see your room? It's the east chamber. Go up the stairs and turn to the right; it's the second door."

"Thanks," said Sidney. "I would like to get rid of the dust a bit."

He went up the dusky oak stair. The house was carefully darkened to keep out the heat and to discourage the flies. He found his room easily. His trunk was there. The air was fragrant with the perfume from a nosegay of sweet peas and mignonette which stood in a willow-pattern bowl of old blue. Associating each gracious deed with *her* gracious presence, he said to himself:

"Vashti—Queen Vashti—has been here." Then he murmured to himself, "Vashti!"

> "The first sweet name that led
> Him down Love's ways."

When he descended in flannels a little later, he found the two girls waiting on the porch. Vashti was sitting on the steps. Mabella was leading a long-suffering cat up and down by its forepaws, a mode of progression which evidently did not please the cat, whose tail switched viciously at each step. It was released as Sidney stepped out of the hall, and relieved its feelings by deliberately walking over and scratching the old collie's nose, as he lay sleepily waiting for the signal to start. The collie, rudely awakened from his dream, sneezed and turned an appealing look at Mabella, who caught him by his feathery ears and expressed her sympathy in words somewhat unintelligible to the human intellect, but evidently well understood by Bunker.

"Don't forget them cups," called Temperance after them. "And don't spill all that milk afore you get there. It won't make the crops grow." Then she betook herself indoors, to muse upon the advisability of making hot biscuits for supper, and to commune with herself upon the absurdity of men who wore flannel trousers.

"My soul!" she said, in recounting the experience to one of her neighbours, "it gave me a turn when I saw him in them white things. First off, I says, 'He's forgot to dress himself.' Then I saw they was white trousers. Poor crittur! He needs something to set himself off; he's poor looking alongside of Lanty."

But Miss Tribbey's judgment was not to be trusted in respect to masculine good looks, her one unit of comparison being yellow-haired Lanty Lansing, who, tall, broad-shouldered and straight-limbed, was a man among a thousand. Sidney Martin had his fair share of good looks. Under any circumstances it would have been impossible to take him for anything but a gentleman, a gentleman by breeding, education, and natural taste. He, too, was tall like Lanty, but much more slender. He had grey eyes—the dreamy eyes of Endymion, slender, nervous hands, and graceful gestures. He walked with something of a scholar's stoop, and had the pallor of the student. Above all, his face was irradiate with kindliness towards every living thing. His eyes had the dilating pupils of those who are dreamers of dreams. It might be that the ideal would take him greater lengths than the truth. About his mouth lay always a touch of pity—pity for the world about him, which, to his eyes, was so blind to the true good, so bent upon burdening itself with baleful creeds which disintegrated the universal brotherhood of man.

The three young people, escorted by the collie, left the house, and turning away from the road, proceeded along a lane which was really a continuation of the avenue without the grateful shade of the trees. The dusty way was strewn with fragrant hay which had fallen from the waggons on their way to the barns. They passed the two broad, shallow ponds, overgrown, as Miss Tribbey had said, with water-weeds and bulrushes. Only a shallow, unwholesome little pool of water remained in each; thirsty birds fluttered about the margin, and, as the three passed, the frogs plunged into the water from every side. The collie walked sedately into the middle of one of the pools, then came and shook himself beside Mabella, spattering her skirts.

The heat was breathless; the earth, beneath the inquisition of the sun, suffered but was mute. And presently they saw the hay-makers, the two sweating horses in the mower, the man tossing the windrows into coils. A great oak tree stood solitary in one corner, and thither the girls directed their steps; a brown earthen jug of water, covered by the men's coats, stood in its shadow. Mabella took off her sunbonnet and waved it wildly by one string. One of the men sent back an answering shout, and tossed a forkful of hay into the air. The sun glinted from the burnished steel of the fork to the yellow hair of its wielder.

"That's Lanty," said Mabella to Sidney, with a certain shy personal pride in her accent.

"Our cousin, Lansing Lansing," amended Vashti.

"Does he live with you?" asked Sidney.

"No! Oh, no! He has a farm of his own, but his haying is all done, and he has come over to help Dad."

"The farmers help each other here, when they can," said Mabella.

Sidney felt enthusiasm surge within his breast; was not this practical communism?

The men had left their work and were coming towards them.

"That's Nathan Peck," said Mabella, "on the left."

Sidney saw him; a serious, sunburnt man, with mild, light-coloured eyes and straight, straggly hair. He was very thin, and wore a woollen muffler around his neck.

"Do you see that scarf? Temperance gave him that three years ago; he's never been seen without it since." Mabella whispered this hastily to Sidney.

"Warm devotion, isn't it?" inquired Sidney as he rose to go and meet his host.

"Isn't he fun?" asked Mabella of Vashti.

"It all depends on taste," said Vashti, indifferently. Mabella did not hear her. She was gazing at her cousin Lanty as he came towards her some yards in advance of the others. Clad in blue jeans, with his shirt open at the throat and his sleeves rolled up to the elbow, Lanty was a man to win ninety-nine women out of a hundred. The odd woman would see, perhaps, too great a capacity for enjoyment in his face; too little of self restraint, too much generosity, too little cool judgment; but if she were discerning enough, she might pierce yet deeper to that natural nobility of character which, through miry places and sloughs of despond, would yet triumphantly set Lanty Lansing upon the solid rock of men's respect.

"Well—you're a sight for sore eyes," he said, flinging himself at the feet of his cousins. "It's worth working for to get over to the shade—and you."

His first words had seemed to address Mabella; his glance took in both his cousins, and each girl took the meaning of the words home to her heart, and doled out a niggard portion to the other. Mabella's confidence had given place to a shy eagerness to please the man she loved. Her eyes dwelt upon him, eager to catch each glance, and she felt that as often as his eyes lighted upon her an unconscious tenderness deepened his voice.

The situation was perfectly apparent to Sidney when he arrived with old Lansing a moment later. Yet Vashti Lansing's blinded eyes saw nothing of it. Rapt in a superb egotism, she erred much in underestimating her fellows. A more dangerous thing, perhaps, than to over-estimate ourselves. Some instinct made her aware of the splendour of her form; besides that, the women of her race had all been mageful creatures. She had an unfaltering belief in the potentiality of her own will. Long ago they had burned one of her forbears as a witch-woman. They said she caused her spirit to enter into her victims and commit crimes, crimes which were naively calculated to tend to the worldly advantage of the witch. Vashti thought of her martyred ancestress often; she herself sometimes felt a weird sensation as of illimitable will power, as of an intelligence apart from her normal mind, an intelligence which wormed out the secrets of those about her, and made the fixed regard of her large full eyes terrible. The film of vanity dimmed them somewhat, but when some rude hand should rend that veil away, their regard might be blasting.

Lanty's wide hat was cast with apparent carelessness upon the grass between him and Mabella; their fingers were interlaced beneath it, or, rather, Mabella's trembling fingers nestled in Lanty's palm. He held them tighter and tighter. A little tremor from her heart communicated with his heart as the electric spark traverses the cable. At the same instant they looked at each other, and read life's meanings in each other's eyes. For the moment—unfaltering, steadfast, penetrating—blue eyes met the blue.

There was the pause of a heart-beat. Then Mabella's filmed with sudden self-consciousness, and triumph lighted the man's bolder eyes. Mabella almost wrenched her hand free and raised it involuntarily to where her heart, grown too great with its treasure of love, throbbed heavily. Lanty rose to his feet, bareheaded in the sunshine, blinded by the glory and promise of the love he had seen in those kindred eyes. He stood for a moment looking down at her; she looked back at him. Her lips were tremulous, but there was an appealing trust in their sweetness. Lanty could not trust himself farther.

"I'll be off to my hay," he said in vibrant tones. "I hope to see a great deal of you," he added, turning to Sidney. "You must come over and see me; whenever you want a horse to ride, there's one at your disposal. Good-bye, girls, till supper time. Good-bye, Mabella." She looked at him, and he went off to his work, scarce believing in his own happiness, seeing all golden about him, all fair before him—and this passed amid a group of people, one at least of whom should have had sharp eyes.

One person indeed had noted all—Nathan Peck's light eyes were eloquent of mute sympathy. He, good soul, loved bustling Temperance Tribbey with all his being. Whilst Lanty and Mabella had rested with their hands clasped beneath the old wide hat, Nathan's gnarled fingers had caressed the ends of his muffler. Temperance was always and invariably right, that went without saying, and yet—nineteen years!—surely she was a little hard on them both? Nathan rose with something like a sigh, and proceeded to his work thoughtfully. Sidney talked to Mr. Lansing and feasted his eyes on the suave grace of his daughter. Mabella, her heart too full for careless speech, rose, and, under pretence of chasing the collie, contrived to start down the lane alone. As she reached the bend which would hide her from Lanty, she turned. He was leaning upon his fork, gazing after her. She waved her hand swiftly to him, then turned abruptly and proceeded upon her way, a demure little figure in her pink sunbonnet.

Life stretched before her in a new aspect; the gate was opened, but the way was unfamiliar, and her feet faltered before it. She arrived home very soon, and sought Temperance in the kitchen.

Temperance was watering her geraniums in the window, and thinking a small kitchen of her own would be more cosy than the great kitchen of Lansing House.

"Temperance," said Mabella, catching hold of a corner of Miss Tribbey's apron, "Temperance, you weren't cross this afternoon when I pulled you about?"

Miss Temperance looked at her, and set down the old tea-pot which she used as a watering-can.

"Say?" insisted Mabella, pleating up the corner of the apron.

"What ails the child?" said Temperance—a sudden memory of Mabella's childhood coming to her, again she saw her a yellow-haired baby with irresistible ways.

"But did you mind?" asked Mabella, her lips beginning to quiver.

"Bless it! No, indeed. My lamb, what kind of a cross old stick do you think I am?"

"Temperance, are you *very* fond of Nathan?"

"My soul!" said Temperance. "What next—Nathan?"

"Because you ought to be if you're not," said Mabella. "Oh, you ought to be. When a person cares about one you ought to love them—*love* them with all your soul. It's so little to give in return; so——" and then Mabella was in Miss Tribbey's arms, crying as if her heart would break.

And blustering Miss Tribbey petted her and quieted her, and got her out of the way before Vashti and Sidney entered with the dishes from the field, taking her upstairs and putting her to bed as she had done long before when Mabella was a little motherless baby.

"You lay still there," said Temperance, pausing by the door. "You lay still there, and I'll fetch up your tea."

"You're a dear," said Mabella with a catch in her voice.

Miss Tribbey departed. Wise in her kind old fashion she asked no questions. Miss Tribbey had been young in years like Mabella once, and her heart was young yet.

"Pore girl!" said Temperance to herself, resuming the watering of her geraniums. "Pore Mabella! She ain't got no mother."

Perhaps all the dew which fell upon the geraniums did not come from the old tea-pot. Miss Tribbey's mother had been alive when lanky Nathan Peck began "keeping company" with Temperance. Up-stairs in a certain box there yet were quilts that she had "patched" in anticipation of the wedding which Miss Tribbey's sense of duty had deferred all these years.

Miss Tribbey sighed, and went and carefully considered her countenance in the little square of greenish glass which served as a mirror in her kitchen. She turned away with something like a sob in her throat. "I'm losin' my looks," she said. Then after a moment's pause she drew herself a little more erect, and going to a drawer put on a huge and fresh white apron. She was meeting the ravages of Time with the defences at her disposal. Brave Temperance!

When some two hours later Nathan Peck entered for his supper with the others, he thought that never, surely in all the world, could there be a more soul-satisfying sight than his "Temp'rins."

"She beats all the young 'uns yet, by jing, if she don't!" he said to himself as he soused himself with soap and water by the door before he came in.

"Here's the comb, Nat," said Temperance, handing him that useful article. He took it, combed his straggly hair straight down over his eyes, and then looked at Temperance appealingly through the ragged fringe.

Temperance's heart was very soft to-night. She took the comb and parted his hair. When she had finished, she let the palm of her hand smooth over the top and rest an instant. He caught it, and the two looked at each other. What were years and hard-wrought hands to them? They saw themselves young and beautiful in each other's eyes. That sufficed them.

Meanwhile Lanty had passed through the kitchen to the front porch, and not finding Mabella there had come back to the kitchen.

"Well, Temperance," he said cheerily, "how's the world using you? And Temperance—where's Mabella?"

"She's layin' down," said Temperance; "she had a sort o' spell when she came in and I made her go to bed."

"What kind of a spell?" demanded Lanty, his heart standing still.

"Nerves," said Miss Tribbey briefly, avoiding the anxious blue eyes of her favourite. She did not know how far matters had gone, nor how clear an understanding there was between the young people. Miss Tribbey was too staunch a woman to betray her sex even in a good cause (and the making of a match between these two Miss Tribbey regarded as a distinctly good cause).

"Is it—is it her head?" asked Lanty miserably.

Miss Temperance eyed him severely—but she had misjudged her own strength.

"It's jist nothin' but nerves," she said—"girls' nerves; they're naterally nervous, girls is, and M'bella ain't one of your coarse-grained sort. She's easily upset and tender-hearted as a chicken. My soul! how all the brute beasts love her and how she sets store by them. I tell you that girl can't pass a hen without sayin' something pleasant to it. She'll be all right to-morrow;

but Lanty"—she quickened her speech as they heard steps coming to the kitchen—"Lanty, she's got no mother."

Lanty caught her hand—"I'll be everything to her, if she'll let me," he said.

Then the others came in. Vashti, her father, and Sidney from the porch, and Nathan from the back doorsteps, where he had been hugging his happiness by himself.

"Where's M'bella?" demanded her uncle as they sat down. Vashti looked at Temperance for the answer.

"She's layin' down—got a headache with the heat."

Nathan looked up with such sparkling intelligence that Miss Tribbey was forced to reduce him by a look. So he obliterated all expression from his face and fell to his supper with a gusto.

"Well, I declare," said the old man; "she must be terrible bad if she couldn't stay up for Mr. Martin's first meal with us."

"Oh, you mustn't mind me," said Sidney hastily, "and I do wish you would call me something a little more familiar than 'Mr. Martin.' My father always called me Sid."

"Sid you are, then," said old Lansing heartily; "it's mighty handy, that name. If there's anything I hate it's a name a mile long. Nothing like a short name for a dog or a person, I say. For horses and sich it don't matter much, but when you want t' call a dog there's nothing like a good plain name." The old man ran on garrulously, now and then arresting himself to say the others were quiet. Considering that their quietude was somewhat compulsory, as he talked all the time, it was rather astonishing he found it food for comment.

"Well—M'bella do miss considerable," he said; "she's always got something to say, M'bella has. Sometimes 'taint over-wise, but it's always well-meaning. M'bella ain't one of your bristle-tongued women. I tell you I've known women with rougher tongues than a cat's."

"Men's tongues is a good deal like dogs', I notice," said Miss Temperance scathingly—"that long they can't keep 'em between their teeth. Mighty loose hung, men's tongues is."

"When is the Special Meeting, father?" asked Vashti. Sidney thought how gratefully her soft voice sounded across the strident tones of her father and Temperance.

"Wednesday night," he answered. "You'll go, Sid? And you'll be there, Lansing?" The last words were spoken in a tone which challenged denial. But Lanty was in a mood of Quakerish peace. He simply nodded. Old Lansing looked very pleased.

"Special meeting!" said Sidney. "What for? What sort of meeting?"

"To pray for rain," said Vashti. "If we do not have rain, the poor people will be ruined and all of us will suffer. Already the hay is lost; we should have had the meeting earlier."

"Then you think—you believe—you believe the meeting will do good?"

"I believe in the answer to prayer," she said a little coldly; "my father is senior deacon in the church."

This seemed hardly a reason for her personal beliefs, but Sidney did not say so.

He began to see her in a new light—a noble daughter of a tottering faith. And as one admires the devotion of a daughter to an unworthy parent, so he admired Vashti in this guise also. The loyalty which made her blind to the faults of a creed was perhaps more admirable than a clearer vision which would have made her a renegade to the faith of her fathers. So Sidney Martin thought as they sat out on the front porch, watching the fireflies flitting in the darkness, living sparks of light, and listening to the cadence of Lanty's violin as he played snatches of old love songs, putting his heart into them—for a little time before he had heard a window softly raised, and he knew that Mabella, too timorous to meet him face to face yet, was listening to and drinking in the message of his music.

CHAPTER III

"You never can wash your hands clean in dirty water," said Temperance to Nathan, "no more'n you can wash a floor with a dirty mop. Throw dirt and the wind'll carry it back in your own eyes. You can't splash mud without gettin' spattered yourself." Thereupon Temperance rattled her dishes violently with an energy almost offensive. Her remarks were in the nature of a parable intended to impress upon her admirer her superiority to, and contempt for, ill-natured gossip.

Nathan bowed his head to the blast, waited till the noisy agitation in the dishpan had subsided a little, and then continued to disburden himself of the news he had gathered during the two days which had elapsed since he had seen Temperance.

"Mrs. Snyder has been took again, I saw Sam and he says she's real miserable."

"You don't say!" said Temperance, fairly interested now. "She has a sight of sickness."

"Well, she was took down three days ago," said Nathan, repeating himself. Having no details to give, he uttered this remark with emphasis, as of one giving forth a brand new idea.

"It's just a year ago this very month since she was took down before," went on Miss Tribbey, uttering her reflection aloud as she was wont to do when she had only the cat for auditor. "I remember particularly well because I was making currant jell at the time, and Mame Settle was here and she was helping dish it out, and she burned her hand, and she said she was goin' to set up with Mrs. Snyder that night, and she said she wouldn't get drowsy with that hand keepin' her company. Yes, 'twas this very month."

Temperance having successfully proved her proposition in regard to the date of Mrs. Snyder's former illness, returned with renewed vim to her dishes.

"It's curious how disease comes back," said Nathan reflectively. "There's my grandfather, he died two years before the church was opened, and he had quinsy regular every spring, and Aunt Maria had her erysipelas in March every year regular as sugar making, and old Joseph Muir had his strokes always in July. I can mind that well, his funeral came just in hayin', for it rained terrible when we was comin' back from the buryin' and someone said, 'Lucky is the corpse that the rain rains on,' and old Ab. Ranger said he guessed luck didn't cut much figger with a corpse anyhow, and for his part he'd a sight ruther had his hay dry in the barn as wet in the field. It seemed kind of unfeelin'."

Nathan rose to throw out the dish water for Temperance, a gallantry he always permitted himself when he spent the evening with her. So anxious was he not to miss this pleasure that he usually made a number of false starts, drawing upon himself a kindly rebuke for fidgeting "like a hen with its head off." Nevertheless Temperance secretly counted upon this bit of attention as much as Nathan did. He was returning with the empty pan when suddenly he stopped.

"Gee!" he said, a strong word giving evidence of excitement. "I clean forgot to tell you the news. Len Simpson's dead." Temperance sat down heavily in a chair.

"My soul!" she said. Nathan continued with oratorical importance, feeling that for once he had made a hit.

"Yes, we was puttin' up petitions in Mrs. Didymus's hen house to-day. She's gone cracked on fancy chickings and keepin' the breeds separate and sich nonsense, and we was petitionin' it off and the bound girl said Mister Didymus had been called over to Simpson's terrible suddent, and he stayed to dinner, and he writ a telegraph and sent it off by young Len to Brixton. He died in Boston, and I don't know if the telegraph was to send home The Body or not. But anyhow, Mister Didymus was terrible affected."

"And so he ought to be, remembering all things," said Temperance. "Poor Len—well, when he was keepin' company with Martha Didymus I thought he was the only young fellar I ever saw that could hold a candle to Lanty. Well, well, and Martha's been dead and gone these three years. Pore Mart, died of heart-break, I always said, and so Len's dead in Bosting! What was he doin' there?"

"They say," said Nathan, telling the tidings shamefacedly, as became their import. "They say he was play actin'."

"Oh, pore Len," said Temperance. "To fall to that! And I've heard many a one say that there never was a man far or near could draw as straight a furrow as Len nor build a better stack. Play actin'!"

Just then Mr. Lansing came out to the kitchen.

"It's most time to start," he said. "We'll take the democrat—comin' to help hook up, Nat?"

Nathan followed him to the stables.

Temperance went to get ready for the prayer-meeting for rain.

The two girls and Sidney were sitting on the grass in the sweet, old-fashioned garden, where verbenas elbowed sweet clover, and sweet peas climbed over and weighed down the homely Provence roses, where mignonette grew self-sown in the sandy paths and marigolds lifted saucy faces to the sun unbidden; where in one corner grew marjoram and thyme and peppergrass, lemon balm, spearmint and rue. The far-away parents of these plants had shed their seed in old grange gardens in England. The Lansings had long ago left their country for conscience sake, bravely making the bitter choice between Faith and Fatherland.

The three young people, waiting in the delicious drowsiness of the summer twilight, were environed in an atmosphere of suppressed but electrical emotion.

Sidney Martin felt within him all the eagerness of first love. Every faculty of his delicate, emotional temperament was tense with the delight of the Vision given to his eyes. How could he ever dream that the moths of the mind would fray its fabric or the sharp teeth of disillusion tear it? And indeed for him it remained for ever splendid with the golden broideries of his loving imagination. Vashti dreamed—even as the mighty sibyls of old brooded over their dreams, conscious of their beauty, and filled with the desire to see them accomplished—finding her visions trebly precious because they were her very own, the offspring of her own heart, the begetting of her own brain, the desire of her own will.

She knew that Lanty did not love her passionately, but to this strange woman there was an added charm in the thought that she must do battle for the love she craved. Her whole soul rose to the combat, which she might have gained had she not made a fatal error in overlooking the real issue, which was not to make Lanty love her, but to make him cease loving Mabella.

Mabella's face, in the soft dusk, wore an exalted expression of purity and tremulous happiness. There were soft shadows beneath her eyes, and her hands trembled as she plucked a flower to fragments. Her hidden happiness had so winged her spirit that her slight body was sorely tired by its eagerness. She started at each sound, and smiled at nothing. Sweet Mabella Lansing did not dream that these eyes of hers had already betrayed her precious secret, but they had been read by a kindly heart. Sidney Martin thought he never in his life had seen anything so sweet as this girl's face, lit by the first illumination of love's torch. An epicure in the senses, he realized keenly the delicacy of this phase of young life—like the velvet sheen upon a flower freshly unfolded, like the bloom upon the grape, like the down upon a butterfly's wing, lovely, but destroyed by a touch. Beneath this evanescent charm he knew there was deep, true feeling, but he sighed to think that the world might mar its unconsciousness.

Sidney Martin had no place in his musings for God, yet in the face of Mabella Lansing he saw a purity, a love, a look of young delight so holy, that almost he was persuaded to think of a Divinity beyond that of human nature. But he said to himself, "After all how sweet a thing human nature is; how cruel to seek to believe in that ancient smirch, called original sin. Has sin part or place in this girl, or in Vashti, Queen Vashti, with the marvellous eyes and the splendid calm presence? Vashti, who looks at life so calmly, so benignly——" and so on, for begin where he would, his thoughts reverted to Vashti. She was first and last with him for ever. The Alpha and Omega of his life.

But these things were all inarticulate, and in the old scented garden the three talked of other things. The girls were telling Sidney the story of the Lansing Legacy.

Long, long before, when the Lansings were by far the most numerous family in the country side, when a Lansing preached in the church, when a Lansing taught in the little school, where Lansing children outnumbered all the others put together, the *doyen* of the family was a quaint old man—Abel Lansing. He was very old, a living link between the generations, and spoke, as one having authority, of the days of old. Although a bachelor, he was yet patriarchal in his rule over the wide family connection, and they brought him their disputes to be adjusted, and came to him to be consoled in their griefs. When they were prosperous, he preserved their humility by reminding them of the case of Jeshurun, "who waxed fat and kicked," and the dire results of that conduct; when they complained of poverty or hardship, he told them they should be thankful for the mercies vouchsafed to them, contrasting their lot with that of their fathers, who threshed their scanty crops with a flail upon the ice, in lieu of a threshing floor, carried guns as well as bibles in church, and ate their hearts out yearning for the far-off hedges of England when they had not yet grown to love their sombre hills of refuge.

He was very eloquent, evidently both with God and man. It was his prayer, so tradition said, which brought the great black frost to an end, and it was a prayer of his, addressed to human ears, which stayed the hand of vengeance, when uplifted against captive Indians. How excusable vengeance would have been in this case, and how well mercy was repaid, is known to all who have read of the troublous times of old.

In fullness of years, old Abel Lansing died, and dying, left all he had to the poor of the parish, save and excepting a hoard of broad Spanish pieces. How he had come by these dollars no one knew. The commonly accepted idea was that they had been brought from England by the first Lansing, and kept sacredly in case of some great need. Be that as it may, there they were, stored in the drawer of the old oak coffer which had been made in England by hands long dead.

And Abel Lansing's will directed that to each Lansing there should be given one piece, and in the quaint phraseology of the times, Abel had set down the conditions of his gift. The recipients were bidden to guard the coin zealously and never to part with it save *in extremis*—to buy bread, save life or defend the Faith.

And strangely enough, when the money was portioned out, it was found that for each broad silver piece there was a Lansing, and for each Lansing a broad silver piece. No more and no less. And the country folk, hardly yet divorced from belief in the black art, with the unholy smoke of the burned witches still stinging their eyes, looked at each other curiously when they spoke of the circumstances.

Oh, what an eloquent human history might be written out, if the tale of each of these coins was known! What an encyclopædia of human joys and sorrows! For no Lansing lightly parted with his Spanish dollar, upon the possession of which the luck of the Lansings depended. They were exchanged as gages of love between Lansing lovers. They were given Lansing babies to "bite on," when they began cutting their teeth. They had been laid upon dead eyes. They had been saved from burning houses at the peril of life. And dead hands had been unclosed to show one held clasped even in the death pang.

Vashti drew hers from her pocket, and showed it to Sidney.

Mabella took hers from a little leather bag which hung about her neck. When Mabella's mother had died in want and penury, she had given her three year old baby the piece and told her to hold it fast and show it to Uncle when he came, for at last the brother had consented to see his sister. He was late in yielding his stubborn will, but when once he was on the road a fury of haste possessed him to see the sister from whom he had parted in anger. But his haste perhaps defeated itself, and perhaps Fate, which is always ironic, wished to add another ingredient to the bitter cup old Lansing had been at such pains to prepare for his own lips. His harness broke, his horse fell lame by the way, the clouds came down, and the mists rose from the earth and befogged him, and when he finally arrived at the bleak little house it was to find his sister dead, and a yellow-haired baby, who tottered still in her walk, but yet had baby wisdom enough to give him the shining silver piece and say "from Mudder." Lansing looked at the baby, and at the coin in his hand, and passed through the open door where an inert head as yellow as the baby's lay upon the pillow. He had come tardily with forgiveness; he had arrived to find his sister dead, and to be offered the symbol of the Lansing luck by an orphan child.

Well—that was but one of the Lansing dollars.

Of all old Abel Lansing's hoard there remained but four pieces—of all that family which had possessed almost tribal dignity there were only four left.

"Are you ready?" shouted old Lansing.

The three young people went round to where the democrat wagon stood with its two big bays. Nathan and Temperance stood beside the horse block; as they appeared Temperance climbed nimbly into the back seat, and Nathan, adorned as usual with his muffler, placed himself in front; the two girls joined Temperance, and Sidney mounted beside Mr. Lansing and Nathan. So they set out, leaving the old house solitary in the deepening night.

As they drove along the country road the burnt odours of the dried up herbage came to them, giving even in the dark a hint of the need for rain.

"Has Nathan told you the news?" asked Temperance of Mr. Lansing. "Len Simpson's dead."

"Oh, Temperance!" said Mabella.

"Where—when?" said Vashti.

Temperance was silent, and Nathan, in the manner of those who have greatness thrust upon them, recommenced his parable.

"Oh, poor Len!" said Mabella, wiping her eyes.

"It's very sad for his people," said Vashti. "First to be disgraced by him, and then to hear of his death like this—well—he was a bad lot."

"Oh, Vashti," said Mabella, passionately, "how can you? And him just dead. His mother'll be heartbroken."

"I did not say anything but what everybody knows," said Vashti, coldly. "He drank, didn't he? And he broke Mart Didymus's heart? I thought you were fond of her? It's true he's dead, but we've all got to die; he should have remembered while he was living that he had to die some day. I don't believe in making saints of people after they're dead. Let them live well and they'll die well, and people will speak well of them."

"That depends," said Temperance with a snort. "Some people ain't given to speak well of their neighbours living or dead."

"And some people," said Vashti coolly, "speak too much, and too often always."

"Hold your peace," said her father sternly. "Did you say The Body was being brought home?" he asked Nathan.

"Yes, or leastways, that's the idea, but no one knows for certain."

"Lanty will take it terribly hard," said the old man musingly. "He and Len Simpson ran together always till Len went off, and Lanty never took up with anyone else like he did with Len."

Sidney had been a little chilled by Vashti's attitude towards the death of this young fellow. But with the persistent delusion of the idealist he did not call it hardness of heart, but "a lofty rectitude of judgment"; himself incapable of pronouncing a hard word against a human being, he yet did not perceive what manner of woman this was. He thought only what severe and lofty standards she must have, how inexorable her acceptance of self-wrought consequences was, and he said to himself that he must purify himself as by fire, ere he dared approach the altar of her lips.

Old Mr. Lansing mused aloud upon Len, and his family, and his death.

"Well," he said, "poor Len was always his own worst enemy. Did you hear if he was reconciled before he died?"

"Reconciled," ah surely, surely that is the word; not converted, nor regenerated, nor saved, but reconciled—reconciled to the great purposes of Nature, to the great intention of the Maker; so infinitely good beside our petty hopes of personal salvation. Reconciled to that mighty law which "sweetly and strongly ordereth all things." Reconciled to give our earthly bodies back to mother earth, our spirits back to the Universal Bosom; to render the Eternal Purpose stronger by the atom of our personal will.

The church to which they were going, and which was even now in sight, was a large frame building, whose grey, weather-beaten walls were clouded by darker stains of moisture and moss. Virginia creeper garlanded the porch wherein the worshippers put off their coats, their smiles and, so far as might be, the old Adam, before entering the church proper. Tall elms overshadowed the roof, their lowest branches scraping eerily across the shingles with every breath of wind, a sound which, in a mind properly attuned to spiritual things, might easily typify the tooth and nail methods of the Devil in his assault upon holy things. Indeed the weird sighings and scrapings of these trees had had their share in hastening sinners to the anxious seat, and in precipitating those already there to deeper depths of penitential fear.

Behind the church, in decent array, the modest tomb-stones of God's acre were marshalled. What a nucleus of human emotion is such a church—with its living within and its dead without, like children clustered about the skirts of their mother. Surely, surely, it is, at least, a beautiful thing, this "sure and certain hope of a glorious resurrection"—the hope which had sustained so many weary old hearts in this congregation, when one after another their loved ones went from them to be cradled in Mother Earth!

Well—Religion they say has grown too scant a robe for human reason. Through its rents is seen the glorious nakedness of science; yet surely the strongest of us must feel a tender reverence for the faith typified by such a church as this. The home of simple faith, where simple folk found peace.

In sect this Church was one of those independent bodies of which there are so many in America, which having retained the severe rectitude of the Puritans are yet leavened with evangelical tenderness, and vivified by evangelical zeal. It approximated perhaps more closely to the Congregational Body than any other, and was self-governing and self-sustaining. As the Chicago people date everything from "The Fire" so Dole people dated all their reminiscences from the "Opening" of the "Church," which meant the dedication of the present church, which, in old Mr. Lansing's boyhood, had replaced the humbler log building of earlier days. The minister was chosen for life, and was by far the most important personage of the community. No one disputed his pre-eminence, and public opinion was moulded by his mind. The ministers tilled their gardens, lived simply as their fellows, and beyond a black coat on Sunday, wore no insignia of office; yet that office wrapped them in a mantle of distinction. There was no laughing at holy things in Dole. No Dole children heard the minister and his sermons criticized. The shadow of the great Unseen rested above the humble church and hallowed it.

Mr. Didymus was an old, old man, and his white-headed wife was bowed and frail. The death of their only daughter, Martha, had been a bitter blow. Outwardly they strove to manifest the resignation of God's anointed. At night when they sat alone they held each other's hands, and wept over the bits of needlework the girl had left.

Deacon Simpson was a stern and upright man. No one recognised more clearly than he, that his son Len was no fit mate for brown-haired Martha Didymus. And yet, he loved his boy.

The two young people accepted the judgment upon them. Len's sullen acceptance of the inevitable was broken by fits of hot-headed rebellion against the decorum of the community, which evidently regarded this bitter dispensation as his just due, yet he never gave up hope until pale Martha

Didymus told him to go his way. Then indeed he departed upon his solitary road, and an evil one it seemed to village eyes.

Poor Martha! Duty may excite one to an excess of courage, but it cannot sustain. She "peaked and pined," and the end of it for her was that she was overtaken by sleep before her time, and went to take her place in the silent congregation.

"Ask Mr. Didymus about Len," said Vashti to her father, catching his sleeve, and detaining him for a moment, as he was about to lead the horses into the sheds.

"Yes—if I have a chance," said her father, and he raised his voice to speak to young Ranger.

"Well, Ab, what hev' you been doin' to-day?"

"Hoeing," said the shock-headed young chap laconically.

"Well," said Mr. Lansing approvingly, "it's about all one can do for the roots in weather like this, and a good thing it is too. You know the old sayin', 'You can draw more water with a hoe than with a bucket.' That's true, 'specially when the wells are all dry."

The two moved away together and Vashti turned to the others. Temperance had left to talk to Sue Winder, one of her great cronies. Lanty had joined Mabella and Sidney.

"I'm glad to see you here, Lanty."

The full diapason of Vashti's voice made the little phrase beautiful. It seemed to Sidney she was like some heavenly hostess bidding wanderers welcome to holy places.

"You have heard of poor Len?"

"Yes, ill news flies fast," he said. His brows were knit by honest pain; and regret, which manlike he strove to hide, made his eyes sombre.

"Are they bringing him home?"

"Yes, Mr. Simpson left for Boston by the six o'clock train from Brixton."

Despite himself Lanty's lips quivered. Mabella ventured in the dusk to touch his hand comfortingly. Her intuitional tenderness was revealed in the simple gesture. He looked at her, unveiling the sadness of his soul to her eyes, and in her answering look he saw comprehension and consolation. As if by one impulse their eyes sought the corner where the slender white obelisk marked the grave of Martha; and having singled it out, where it stood like an ominous finger-post on love's road, they once again steadfastly regarded each other, each one saying in the heart, "Till death." And another thought came to each. They mourned for Len, but *she* rejoiced. Perhaps it was unorthodox, but these two, in the first tenderness of their unspoken love, felt sure that Len did not enter the dark unwelcomed.

Night was coming swiftly on—a "black-browed night" indeed. The faces of the four young people shone out palely from the environing gloom.

It was a solitary moment. Sidney sighed involuntarily. He felt a little lonely. Regretting almost that he could claim no personal share in the grief for Len. Vashti heard his sigh and looked at him. By a capricious impulse she willed to make him hers—to make him admire her. She smiled—and let her smile die slowly. As a fitful flame glows for a moment making a barren hearth bright ere it gathers itself into the embers again, so this gentle smile changed all the scene to Sidney's eyes. His heart was already captive, but it was now weighted with a heavier shackle.

Vashti Lansing saw clearly the effect of her smile, and a mad impulse came upon her to laugh aloud in triumph. Every now and then she felt within her the throes of an evil dominant will. Such a will as, planted in the breast of sovereigns, makes millions weep. The harsh bell began to jingle. It was time to enter.

"Come to our pew, Lanty," whispered Mabella, softly.

"Yes, dear," he answered, and both blushed; and thus they entered the church.

Vashti walked slowly up the aisle, looking neither to the right hand nor to the left, but seeing all. How white and calm she was, Sidney thought—but often lava lies beneath the snow.

The deacons entered; tall, spare men, stern-faced and unsympathetic they seemed, yet in their hearts they thought of the one of their number who was journeying through the night to where his son lay dead. White haired Mr. Didymus rose in his place and stretched out his tremulous hands above his congregation.

"Let us pray," he said, and after a solemn pause addressed himself to the unseen.

The greater part of the congregation knelt, the deacons stood erect, as did Lanty and Sidney, although a thought crossed the mind of each of the young men that it would have been sweet to kneel beside the woman he loved.

As Sidney looked about him a great pity for these people filled his heart; the kneeling figures appealed to him poignantly; from his point of view they were less like children gathered about a father, than serfs bending beneath a yoke, which was none the less heavy because it was the creation of their own imagination. The shoulders of the kneeling figures had involuntarily fallen into the pose of their daily toil; there was the droop of the ploughman over his plough; of the tiller over the hoe; of the carpenter over his plane. It was as if, even in prayer, they wrought at a hard furrow. And the women's shoulders! What woeful eloquence in these bent forms bowed beneath the dual burden of motherhood and toil. What patient endurance was manifest beneath the uncouth lines of their alpaca and calico dresses!

From the shoulders his gaze fastened upon the pairs of hands clasped upon the pew backs. Such toil-worn hands. It seemed to him the fingers were great in proportion to the palm, as if they wrought always, and received never. Surely he was growing morbid? And then all the latent pathos in the scene gathered in his heart. All the dumb half unconscious endurance about him pleaded to be made articulate; and as one with unbelieving heart may join in a litany with fervent lips, so Sidney strove to second each petition of the long prayer.

Old Mr. Didymus had long been a spiritual ambassador and he was not unskilled in diplomacy. His prayer was a skilful and not inartistic mingling of adoration, petition, compliment and thanks, adroitly expressed in the words of the Sovereign he addressed, or in phrases filched from His inspired ones. And mid their burning sands, and under their blazing skies, these Eastern followers had not failed to appreciate the blessings of rain.

"O, Thou who in the wilderness did rain down the corn of heaven, that Thy children might eat and be filled; Thou who brought streams out of the rock and caused waters to run down like rivers that their thirst might be quenched, and that they might be preserved alive—Thou of whom it was said of old: 'Thou visitest the earth and waterest it; Thou greatly enrichest it with the river of God, which is full of water; Thou waterest the ridges thereof abundantly; Thou settlest the furrows thereof; Thou makest it soft with showers; Thou blessest the springing thereof; Thou crownest the year with Thy goodness'—Hear us! We beseech Thee! Thou causest it to rain on the earth where no man is—on the wilderness wherein is no man—cause it also to rain upon us. Thou causest it to rain alike upon the just and the unjust, let us not hang midway between Thine anger and Thy love. Remember Thy promise to pour water upon him that is thirsty, and floods upon the dry ground. Thou, O God! didst once send 'a plentiful rain whereby Thou didst confirm Thine inheritance when it was weary'; deny us not a like consolation, we faint beneath the hot frown of Thine anger. Let

Thy shadow comfort us! As the thirsty hart panteth for the waterbrooks, we long for Thy blessing. Before man was upon the earth Thou caused a mist to rise up from the earth and watered the whole face of the earth; continue Thy mercy to us, who, sharing Adam's fall are yet heirs to the Redemption. Slay not us in Thine anger, O Lord! Behold, we are athirst! Give Thou us to drink. Are there any vanities of the Gentiles that can cause rain? Or can the Heaven give showers? Art not Thou He—O Lord our God? Therefore we will wait upon Thee, for Thou hast made all these things. When Elijah strove against the sorcerers of Baal didst Thou not hear him? Like unto him we are cast down before Thee. O grant us our prayer! Show to us also the little cloud like a man's hand that comforted the land of Ahab. Grant that we, too, by faith shall hear 'the sound of abundance of rain!'"

He paused. There was a moment of tense silence.

"And Thine shall be the glory, Amen," he faltered forth brokenly. He had no further words; the advocate had pleaded for his cause. He waited the voice of the judge. There followed a longer pause fraught with the emotion of a great need.

Sidney's heart ached for these people; a thousand inarticulate pleas entered the wide gate of his sympathies and demanded utterance at his lips. A sultry breath entered the open window fraught with the odour of parched earth and burnt-up grass. The old priest and his three grey-haired elders, standing amid the kneeling people, seemed to him like brave standards ready to prop up a falling faith till its ruin crushed them, willing sacrifices for the people; they were mute, but their very presence standing thus was eloquent. Surely the God of their Fathers would remember the children of these men who had indeed "given up all and followed Him" out to the western wilderness? Long ago he had led forth His people out of Egypt. They had murmured against Him yet He had not left them to perish in their sins; was the hand that had given water from the rock and corn from Heaven empty now?

Long ago the great progressive miracle of Nature's processes was inaugurated; were the wheels of God's machinery clogged?

A shrill, trembling treble voice rose brokenly. For a few ineloquent phrases it continued, and then died away in sobs expressive of mortal need. It was Tom Shinar's wife; their farm was to be sold at mortgage sale in the autumn. Mary Shinar had gone herself to plead with the lawyer in Brixton through whom the mortgage had been placed. Mary sat on the edge of a chair in an agony of nervousness whilst the perky clerk went in to state her business, and the lawyer came out of his comfortable office and told her they could stay on the farm "till the crop was off the ground," he did not know the terrible irony of his mercy.

In the light of ordinary day Tom Shinar and his wife bore themselves as bravely as possible. Their neighbours asked them questions as to next year's crops to force them to betray what was a secret only by courtesy. All the community knew the facts of the case, and when Tom, forced into a corner by questions, said "he 'lowed he'd be movin' in fall," every man knew what he meant. When Mary, in a like position, said she "reckoned they wouldn't have to bank up the cellar that winter, 'cos Tom was thinking of changing," the women said to each other afterwards: "They're to be sold out in October—Mr. Ellis is takin' the farm."

A mortgage sale is an ordinary enough event, and the prospect of one not so unique as to require dwelling on, but the sight of Mary Shinar's face as she let it fall between her hands after her abortive prayer, decided the fate of Sidney Martin. The sound of a woman's trembling tones was the touch which sent Sidney over the brink of the pit Fate had prepared for him. The last echo of her shrill voice died away—a sob filled the room of the wonted Amen. The sob did not die till it filled Sidney Martin with fatal inspiration; again he agonized in one of his childish visions when the Pain of the world, exaggerated by his morbid mother's teachings, seemed to environ him with the tortures of hell. His supra-sensitive personal atmosphere was surcharged with electrical currents of pain and need and want, defeated effort, dead hope, fruitless battling, and these discharging themselves in his bursting heart, filled it with exquisite agony. His spirit battled against his imagination and rushed to his lips.

He began to speak. No one in that congregation could ever recall one word or phrase of Sidney Martin's prayer for rain. As the "poor, poor dumb mouths" of Cæsar's wounds lent Antony eloquence, so each line and careworn furrow upon the countenances of those about him sped the speech of Sidney Martin.

The women sobbed aloud, the men felt their heavy souls lifted up. Lanty, whose ardent nature made him peculiarly susceptible to the charm of eloquence, fell upon his knees involuntarily. Mabella felt a pleader powerful enough to win their cause was here amid the stricken congregation, and Vashti felt once again a wildly exultant throb of her own power which had won such a man.

Yet—what manner of prayer was this? Herein were no phrasings from Holy Writ; no humble appeals to a pitying Christ, a personal God.

Sidney Martin, standing amid this congregation of orthodox souls, was pouring forth what was neither more nor less than a pantheistic invocation to the Spirits of Nature, bidding them be beneficent; addressing them with Shelleyan adoration, and with as strong a sense of their existence as ever inspired Shelley's immortal verse. And thus within these walls wherein was preached naught but "Christ and Him crucified," Sidney Martin addressed himself to "Nature—all sufficing Power," and did it, moved by no irreverence, stimulated by the same needs which had wrung forth the few pleading words from pious Mary Shinar. And whilst he, in bitterness of spirit, realized afterwards the *grotesquerie* of his action, yet those who were his hearers that night, and for many times afterwards, never saw the great gulf fixed between his adorations and their beliefs. And is it not a hopeful and solemn thing to find the Faith in a living Christ so closely allied to honest reverence for nature? To find Nature so close akin to God that their worshippers may interchange their petitions? It is very significant that—significant as all things are of the immutable and sacred Brotherhood of Man.

Christian, Deist, Buddhist, Atheist, by whatsoever name we choose to call ourselves, we are all bound together by the thongs of human needs and aspirations.

How vain to seek to deny that kinship. How futile to strive to blot out the family resemblance betwixt our prayers and theirs!

For *malgré* himself man prays always. His mere existence is a prayer against the darkness and the chaos of the void.

Sidney's voice rose thrillingly through the tense silence. He had that God-like gift—natural eloquence, and under its spell his hearers forgot in part their woes, and began to take heart of hope whilst he plead with Mother Nature not to be a step-dame to her sons, and besought the "beloved Brotherhood," earth, air and ocean, to withdraw no portion of their wonted bounty.

As his eloquence carried his listeners beyond their fears, it bore himself beyond their ken, till suddenly alight upon the highest pinnacle of thought, he paused to look beyond—hoping to behold

"Yet purer peaks, touched with unearthlier fire,
In sudden prospect virginally new,
But on the lone last height he sighs, 'tis cold,
And clouds shut out the view."

Sidney saw but a misty void peopled with the spectral shapes of his doubts, which gibbered nebulously through the veil at him. Speech died upon his lips. His voice, arrested midway in a phrase, seemed still to ring in the listening ears. It was as if one paused in an impassioned plea, to hear the answer rendered ere the plea was finished.

And the answer came.

A long sighing flaw of wind swept about the church, cool and sweet, and ere it died away rain was falling.

"Amen," said every pair of lips in the church save the pale, quivering lips of Sidney Martin. The coincident arrival of the longed-for blessing added the finishing touch to his nervousness. He rose from the pew into which he had sunk for a moment and swiftly passed down the aisle, hearing, ere he reached the door, the first lines of the hymn of Hallelujah, which went up from the grateful hearts behind him. His whole being revolted against his recent action.

The rain beat down violently; the parched earth seemed to sigh audibly with delight, and within the church all the voices vibrant with justified faith seemed to mock at his depression. He could not explain his action to himself. What explanation then was possible to these simple folk?

Could he say to them—to Vashti—(he named her name in his thoughts, determined not to spare himself). Could he say to Vashti, "I do not believe in your God—nor in the man Christ Jesus, nor in prayer. Yet I stood in the church and asked a blessing. I defiled your fane with unbelieving feet. I do not know why I did it"? It was weak that, certainly! He imagined the scorn in her clear eyes; now eyes in which scorn is so readily imaginable are not the best eyes—but he did not think that. What was he to do? He had been weak. He must now be strong in his weakness.

The church door opened, one and all emerged upon the long verandah-like porch, and gathered round him, shaking hands with him.

"The Spirit indeed filled you this night, Brother," said white-haired Mr. Didymus.

"Yes—you wrestled powerfully," said Mr. Lansing.

"It done me good that prayer of your'n," said Tom Shinar, and the words meant much.

"We have much to thank you for," said Vashti's sweet tones, and for the first time he looked up, and when he met her approving eyes, the garments of his shame clung tighter to him.

Mabella gave him her hand a moment and looked at him shyly.

Lanty stood a little aloof. He was a good young chap with honest impulses and a wholesome life, but he never felt quite at ease with parsons. Lanty placed them on too high a pedestal, and after having placed them there found it strained his neck to let his gaze dwell on them. He had a very humble estimate of his own capacity for religion. He was reverent enough, but he had been known to smile at the peculiarities of pious people, and had once or twice been heard making derogatory comparison betwixt precept and practice as illustrated in the lives of certain potable church members.

"Well," said Temperance energetically to Sue Winder. "Well! I'm sure I never so much as 'spicioned he had the gift of tongues! After them white pants!! He talked real knowin' about the fields and sich, but to home he don't seem to know a mangel-wurzel from a beet, nor beets from carrits."

"There's no tellin'," said Sue, who was somewhat of a mystic in her way. "P'raps 'twas The Power give him knowledge and reason."

"Well—I don't know," said Temperance, "but if he stands with that eavetrough a-runnin' onto him much longer it'll give him rheumatics."

"Temp'rins is powerful worldly," said Sue regretfully to Mary Shinar as Temperance left her side to warn Sidney. Her experienced eyes saw his deathly pallor; she deflected her course towards Mr. Lansing where he stood among the worthies of the congregation giving a rapid *resumé* of Sidney's history so far as he knew it.

Temperance was a privileged person. She broke in upon the conclave with scant excuse.

"Mr. Martin is fair dead beat," she said without preface. "He's got a look on his face for sickness. He'd better be took home. Nat, will you fetch round the demicrit?"

Nat departed. Temperance strode over to Sidney.

"If you'd come in out of the rain you wouldn't get wet," she said, as if she was speaking to a child; "we're goin' home direckly, and there's no good running after rheumatics; they'll catch on to you soon enough and stick in your bones worse nor burrs in your hair."

Sidney moved to the back of the porch and leaned wearily against the church.

"It seems to me he'll get middlin' wet driving home anyhow," said Mr. Lansing.

"Do you think I came to a prayer-meeting for rain without umbrellas?" snorted Temperance. "Them and the waterproofs is under the seats."

There was silence.

A demonstration of faith so profound was not easily gotten over.

Graceless Lanty sniggered aloud.

The listeners felt themselves scandalized.

"Well, I *declare*!" said Mrs. Ranger, openly shocked.

"Did you bring your umbrell and your storm hood?" asked Temperance.

"No," snapped Mrs. Ranger, remembering her new crape.

"That's a pity," said Temperance coolly, "seeing you've got your new bunnit on—when you knew what we came here for."

In the parlance of the village, Temperance and Mrs. Ranger "loved each other like rats and poison."

Nat arrived with the democrat—jubilant over "his Temp'rins'" foresight. "That's what I call Faith," he said, handing out the coverings.

"I'm glad he told us," whispered Lanty to Mabella "if he hadn't—I'd have thought 'twas your waterproofs."

And Mabella, though she was a pious little soul, could not help smiling rosily out of the waterproof hood at her lover's wit, and what with the smile, and the ends of her yellow hair poking out of the dark hood, and her soft chin tilted up to permit of fastening a stubborn button, Lanty had much to do to abstain from sealing her his then and there before all the congregation.

All was at length arranged, and Temperance went off with her party dry beneath the umbrellas. The rest of the congregation took their drenching in good part. They were not going to complain of rain in one while!

CHAPTER IV

The next day dawned with pale rain-bleached skies and fresh sweet odours of reanimated vegetation, but it dawned heavily for Sidney Martin. During the drive home from the church the evening before they had all been somewhat silent.

"Are you studyin' for the ministry?" old Mr. Lansing had asked.

"No—oh no," said Sidney, flushing unseen in the dark.

"It seems like you had a call," said old Lansing, wishing he had not said quite so positively at the church that his visitor was qualifying for the service of God, and certainly from Mr. Lansing's point of view he was justified in his assertion.

Young men in delicate health who could pray as Sidney Martin had prayed seemed to be the real ministerial material.

"Wouldn't you like to be the minister?" asked Vashti.

People in Dole usually employed the definite article in referring to men of the cloth. To the Dole mind it smacked of irreverence to say "a" minister, as if there were herds of them as there is of common clay.

There was a *soupçon* of surprise in Vashti's tones. How quickly the acid of deception permeates the fabric of thought!

"I have no call to the ministry," answered Sidney—employing the slang of the cult glibly to please the woman whom he loved.

"But if you felt you were called you would let nothing stand in your way—would you?"

"No," said Sidney, glad of an opportunity to say an honest word frankly. "No."

There was little else said. When they came to the cross-road Mr. Lansing halted and Nathan Peck got out of the waggon to walk down the Brixton road the quarter of a mile to where he lived with his mother.

He stood a lank ungraceful shape in the gloom.

"Here, Nat," said Temperance, "take my umbrell."

"Not by a jugful," he said. "Why, Temp'rins! you'd be soaked clean through."

"Temperance can come under my umbrella," said Mabella, divining the pleasure it would give Temperance to yield up hers to Nathan.

"I've got my muffler on," said Nathan stoutly.

"Here!" said Temperance, a trifle imperatively. "Good-night, Nat."

The bays pulling at the reins started forward and Nat was left with the umbrella.

"Would you care to offer a few words of thanks for the vouchsafed blessing?" asked Mr. Lansing, with a laudable desire to make his saintly guest comfortable, entering the house after putting the bays in their stable.

"Blessing!" echoed Temperance irascibly. "He's had enough of blessings this night, I'm thinkin'; it's boneset tea he needs now."

"Woman!" said Mr. Lansing. Vashti looked her cold displeasure. The word and the look did not disturb Temperance.

"Lend a hand, M'bella," said she; "we'll go and get them yarbs."

"Oh—thank you, Miss Tribbey," said Sidney, feeling strangely comforted by this motherly old maid's attentions. "But——"

Temperance cut him short, looking at him with grim kindness and heeding his protest not at all.

"Your face is as pale as buttermilk," she began. "Now what you'll do is to go upstairs and go to bed. Mind shut your window down, for rain after a drought is terrible penetratin'. When this boneset tea has drawed Mr. Lansing 'ill bring it up to you."

Mabella was bustling about getting a lamp to go to the garret for the herbs.

"You are very good," Sidney said to her as one might praise a willing child.

"Light heart makes light foot," said Temperance oracularly. Mabella smiled brightly and blushed.

Vashti standing with the dark folds of her cloak slipping down about her superb figure, noted the blush, and connecting it with the eagerness of Mabella's aid to Temperance concluded that Mabella was casting eyes upon Sidney. Vashti's eyes grew deep and sombre. A pale smile curled her sculptural lips; such a smile as Mona Lisa wears in her portraits.

Mabella's coquetries against her power! Bah! a sneer flickered across her countenance, erasing expression from it as acid cleans metal of stain. But she was shaken with silent rage at the mere idea. She let her white lids fall over her full eyes for a moment; then crossed to where Sidney stood. She always seemed to move slowly, because of her long gliding paces, which in reality bore her swiftly forward. She looked into his eyes. "I am so sorry," she said—her voice, always beautiful, seemed to his greedy ears more than exquisite now—

"I am so sorry you are not well. You will go upstairs, won't you, and take what Temperance sends you? You are not suffering?"

Her wonderful eyes seemed wells of womanly concern for him. They searched his as if eager to be assured that there was no other ill troubling him than was apparent. A happy tremor thrilled his heart.

"I shall be quite well, I hope, in the morning," he said. "I have bad headaches sometimes. This is the beginning of one I suppose."

He shivered with cold.

"Ah!" she said, "you must go away at once. I'm afraid you feel worse than you will admit. If it was only your head I might help to cure it; but really you had better go—" she looked at him—was it compellingly or pleadingly? "Go," she half whispered, with obvious entreaty in her eyes; then she veiled it with a smile of mock deprecation, as if—his heart stood still with delight—as if she was loathe to see him go—yet for *his* sake wished it. Temperance and Mabella having been to the garret where the herbs were hung to dry, re-entered the kitchen in time to hear Vashti's good-night words.

"It's a deal easier," said Temperance, in the course of a circumstantial account of the occurrence later on. "It's a deal easier to say 'Go' with a dying-duck expression, turning up the whites of your eyes, than to go yourself up them stairs and that pesky ladder to the garret for yarbs."

Fortunately Sidney never knew of Temperance's profane criticisms upon his goddess.

"Yes—I will go," he said to Vashti. He spoke vaguely, as of one hardly awake to the realities about him; and indeed he was stunned by the glory that suddenly had shone in upon him when her feigned solicitude made his heart leap.

"You are very good," he said.

"Ah, no—" said Vashti simply, but her eyes were eloquent. Girlish coquetries became subtle sorceries as she employed them.

The boneset tea had been duly despatched, but morning found him racked by an intolerable headache, that acme of nervous pain of which only supra-sensitive folk know. He half staggered as he sought the porch.

Temperance came to him presently.

"How do you feel this mornin'?" she asked.

He looked at her, his blood-shot eyes dizzy with pain.

"I'm not over well," he said. "My head——"

"I'll bring your breakfast here," said Temperance and departed. He sat down upon the porch step and leaned against the pillar, the same against which Vashti had stood that night in the after-glow. The thought was pleasant, but it was better to open his eyes and see standing before him, strong and calm, the Queen of his dreams.

"Don't rise," she said. "Is it your head?"

"Yes," he said, half closing his eyes again, for her form seemed to be reeling across his vision. "Yes."

"What do you do for these headaches?" asked Vashti.

"Oh, bromides and endurance," he said.

"Well—wait till you breakfast and I'll try it I can cure it," said Vashti. "Here is Temperance coming."

Temperance and her tray arrived at the moment. Temperance put it down on the step and went down the sandy garden paths whilst he ate, pulling up a weed there, straightening a flower here. Mabella came out to the porch, or rather came and stood in the wide doorway a moment. Mabella had on her pink dress—at that time in the morning! Vashti's eyes grew sombre for an instant; she liked battle, but not presumption, and surely if, from whatever motive, she chose to smile upon Sidney, it was not for Mabella to oppose herself and her charms to *her* will.

Temperance came back for the tray, which she found untouched, save for the tea which Sidney had drunk so eagerly.

"Where is Mr. Lansing?" asked Sidney, as Temperance stood holding the tray under one arm with its edge resting upon her hip. "He will think I am very lazy."

"He's gone over to Brixton to find out when The Body will arrive," said Temperance.

Poor Len! In life he had been "that Len Simpson," and not one of his neighbours would have crossed the threshold to greet him, unless prompted by that curiosity which leads us to pry into the misdeeds of others. Now he was a Body, and more than one of the Dole people had left early like Mr. Lansing upon the odd chance of meeting his corpse at Brixton.

Ah, poor, inconsistent humanity which fills dead hands with flowers and denies eager palms one rose, and doubtless these things must be. Yet we can imagine that a higher race than we might well make mock of our too severe judgments—our uncomprehending judgments, and our tardy tendernesses.

"You will make your passes for Mr. Martin, won't you, Vashti?" said Mabella, "and Temperance and I will see that you are left quiet. Vashti is a witch, you know," she continued to Sidney; "she will steal your headache with the tips of her fingers."

Temperance snorted and entered the house without more ado.

Mabella nodded and smiled and followed her.

"I can't abear them passes and performances," said Temperance to Mabella. "It gives me the shivers. Vashti commenced on me onct when I had neuralgia and I was asettin' there thinkin' when I got better I'd make some new pillars out of the geese feathers, and all at onct Vashti's eyes began to grow bigger and bigger—just like a cat's. They're cat green Vashti's eyes is, call 'em what you like—and her hands apassin' over my forrit was just like cat's paws, afeelin' and afeelin' before it digs its claws in. My! I expected every minnit to feel 'em in my brains, and with it all I was that sleepy. No, for me I'll stick to camfire and sich."

"Who's a silly, Temperance?" demanded Mabella.

"You ain't bridle-wise yet," said Temperance, using her accustomed formula of rebuke. And Mabella laughed aloud in defiance of reproof. The girl's heart sang in her breast, for when Lanty helped her into her waterproof the night before he had whispered—

"At seven to-morrow night in Mullein meadow."

She had smiled consent.

Would this long day never pass?

Vashti and Sidney were thus left solitary upon the shaded porch.

"Can you really cure headaches?" he said.

"We will see," she answered. "But I think you had better sit in that chair." He sat down in the rocking-chair she indicated. It was very low. As she knelt upon the top step before it her head was on a level with his. How beautiful she was, he thought. How divine the strong white column of her throat, exposed down to the little hollow which the French call Love's bed, creased softly by the rings of Venus' necklace.

"I wouldn't think much if I were you," she said, "or at least, not of many things."

"I will think of you," he said, feeling venturesome as an indulged child.

"Ah," she said; "your cure will be quick," and then bending gracefully forward she began making simple strokes across his forehead, letting her finger-tips touch lightly together between his eyebrows, and drawing them softly, as if with a persuasive sweep, to either side. There was much magnetism in that splendid frame of hers, and much potency in her will, and much subtle suggestion in those caressing finger-tips.

"Close your eyes if the light wearies them," she said softly, but he strove to keep them open to catch glimpses of her regal face, between the passages of the hands, so calm in the tensity of its expression. After a little while his eyelids began to weigh heavily upon his eyes.

The grey—or was it green?—orbs watching him flashed between the moving fingers like the sun through bars of ivory. He still watched their gleam intently; seen fitfully thus their radiance grew brighter, brighter, till it blasted vision.

"Close your eyes," he heard a voice say, as from far, far away.

"You will be tired," he muttered, stirring, but his eyes closed. His head fell back against the back of the chair, and strong Vashti Lansing sank back also, pale and trembling.

"Oh!" she said, speaking numbly to herself—"Oh! how long it was. I thought he would never sleep—I," she paused and looked at the sleeping man with pale wrath upon her face; "to think he should have resisted so—I"—she leaned back, worn out, it seemed, and regarded Sidney with venomous, half-closed eyes, and he slept, and sleeping, smiled—for his last thoughts had been of her.

The time which had seemed so long to Vashti had passed like the dream of a moment to him—a dream in which her form had filled the stage of his mind, yet not so completely as to exclude some struggles of the entrapped intelligence against the narcotic of her waving hands. The trained mind by mere mechanical instinct had striven against the encroaching numbness, but Sidney's volition had been consciously passive, and the intelligence left to struggle alone was tangled in the web of dreams. Vashti sat listlessly upon the step for some time—like a sleek, beautiful cat watching a mouse. Then she rose and went within doors to perform her share of the household duties very languidly.

The three women dined alone at twelve o'clock, for Mr. Lansing had not returned, and Sidney still slept. After dinner Vashti disappeared, going to her room and throwing herself heavily upon her old-fashioned couch; she also slept.

Active Temperance fell to her patchwork so soon as her dinner dishes were done, sitting, a comfortable, homely figure, in her calico dress and white apron. Now pursing her lips as she pleated in the seams firmly between her finger and thumb; now relaxing into grim smiles at her thoughts, but always doing with all her might the task in hand.

Mabella essayed her crochet, tried to read, rearranged her hair till her arms ached from holding them up, and found with all these employments the afternoon insupportably long.

About three o'clock in the afternoon Vashti, cool and calm, descended the stairs and went out upon the porch. As she crossed the threshold, Sidney, lying still as she had left him hours before in the low chair, opened his eyes and looked up into her face. She returned the look—neither for a moment spoke. A sudden deep hush seemed to have fallen upon, about them. Had he awakened from his dream, or had she entered it to make the dream world real with her presence? About them was all the shadowy *verdâtre* of trees and vines. Sidney had forgotten where he was—all earthly circumstances faded before the great fact of her presence. He was conscious only that he was Man, and that Woman, glorified and like unto the gods for beauty, stood before him. Were they then gods together?

"Is your head better?" she asked; her full tones did not jar upon the eloquent silence, but her words reminded him that he was mortal.

"I had forgotten it," he said. "I must think before I can tell."

She laughed—just one or two notes fluted forth, but in their cadence was the soul of music. It was as if mirth, self-wrought, bubbled up beneath the dignity of this stately creature, as the living spring laps against the marble basin which surrounds it; and as the tinkle of the spring has more in it than melody, so Vashti Lansing's laughter was instinct with more than amusement. There was in it the thrill of triumph, the timbre of mockery, and the subtlety of invitation.

"Then," she said, "we will take it for granted that it is better. You are like father and the thistles in his fingers. He often tells me how he has been tormented by some thistle, and when I go to take it out, he has to search the fingers of each hand before he can find out where it is. He sometimes cannot even tell which hand it's in."

"Well," said Sidney; "I am like your father. I've lost my head."

"But if it ached," said she; "it was a happy loss."

"I hope it will be a happy loss," he said wistfully.

She smiled gently and let fall her eyelids; no flicker of colour touched her cheeks, nor was there any suggestion of shyness in her countenance. Thus a goddess might veil her eyes that her purposes might not be read until such time as she willed to reveal them.

Mabella heard voices upon the porch and came flying out.

Sidney could not find it in his heart to be impatient with this bright-faced girl, whose heart was so full of tenderness to all living things that little loving syllables crept into her daily speech, and "dear" dropped from her lips as gently and naturally as the petals of a flower fall upon the grass, and as the flower petals brighten for a little the weed at the flower's foot, so Mabella's sweet ways gladdened the hearts of those about her.

"Ah, Mr. Martin," she said, "so you are awake! Was I a true prophet? Yes—I'm sure of it! Vashti's finger-tips did steal the ache, didn't they? They're too clever to be safe with one's purse. But see—have you had anything to eat? No? Why, Vashti," in tones of quick concern, "he must be faint for want of something to eat." She was gone in a moment. With Mabella to know a want was to endeavour to supply it. Ere there was time for further speech between Sidney and Vashti, Temperance had come out. Her shrewd, kindly face banished the last shreds of his dreams. The pearl portal closed upon the fair imageries of his imagination and he awoke, and with his first really waking thought the events of the night before ranged themselves before his mental vision. As he lay awake in the night he had decided that come what may he must put on a bold front before the awkward situation he had created for himself. But if the courage which springs from conscious righteousness is cumulative, the courage which is evolved from the necessities of a false position is self-disintegrating. Sidney felt bitterly that he feared the face of his fellows.

"Eat something," said Temperance, urging the bread and milk upon him; "eat something. When I was took with the M'lary I never shook it off a bit till I begun to eat. It's them citified messes that has spoiled yer stummick. Picks of this and dabs of that, and not knowin' even if it's home-fed pork, or pork that's made its livin' rootin' in snake pastures, that you're eatin'. My soul! It goes agin me to think of it; but there, what kin ye expect from people that eats their dinners as I've heard tell at six o'clock at night?"

Sidney ate his portion humbly whilst Temperance harangued him. He looked up at her, smiling in a way which transfigured his grave, thin face.

"I'm a bother to you, am I not, Miss Tribbey? But it's my bringing up that's responsible for my sins, I assure you. My intentions are good, and I'm sure between your cooking and your kindness I shall be a proverb for fatness before I go away."

"Soft words butter no parsnips," said Temperance with affected indifference. "Fair words won't fill a flour-barrel, nor talking do you as much good as eatin'," with which she marched off greatly delighted. Mabella seeing a chance to tease her, followed:

"If you make eyes at Mr. Martin like that I'll tell Nathan Peck," Sidney heard her say.

"My soul! Mabella, you've no sense, but, mind you, it's true every word I said. I tell you I ain't often in town, but when I am I eat their messes with long teeth."

Sidney moved his camp from the porch to the hammock which was suspended between two apple trees in the corner of the garden. Mabella brought out her sewing, and Vashti her netting, and Sidney spent the remnant of the waning afternoon watching the suave movements of Vashti's arm as, holding her work with one foot, she sent her wooden mesh dexterously into the loops of a hammock such as he was lying in; and at length the shadows lengthened on the grass, and Temperance called that supper was ready.

Mabella Lansing never forgot that repast. It was the passover partaken of whilst she was girded to go forth from girlhood to womanhood, from a paradise of ignorance to the knowledge of good and evil. The anticipation of a new love made these time-tried ones doubly dear. She forgot to eat and dwelt lingeringly upon the faces about her; faces which had shone kindly upon her since she was a little child. The time which had crept so slowly on the dial all day long now seemed to hasten on, as if to some longed-for hour which was to bring a great new blessing in its span.

In retrospect of "the sweet years, the dear and wished-for years," do we not all single out from them one hour becrowned above all others; one hour in marking which the sands of memory's glass run goldenly? Amidst the dead sweetness of buried hours is there not always one whose rose it amaranthine? One, which in the garlands of the past retains ever the perfume of the living flower, shaming the faint scent of dead delights? One hour in which the wings of our spirits touched others and both burst forth in flame? And the chrism of this hour was visible upon the brow of Mabella Lansing. She was sealed as one worthy of initiation into its fateful mysteries. How far away she seemed from those about her; their voices came to her faintly as farewells across the widening strip of water which parts the ship from shore.

"Did you find out about Len Simpson's funeral?" asked Vashti of her father.

"Yes—the buryin's to-morrow, and it seems Len was terrible well thought of amongst the play-actin' folk, and they've sent up a hull load of flowers along with the body, and there's a depitation comin' to-morrow to the buryin' and they say there's considerable money comin' to Len and of course his father'll get it. I don't know if he'll buy that spring medder of Mr. Ellis, or if he'll pay the mortgage on the old place, but anyhow it'll be a big lift to him."

"Why, is it as much as that?" asked Vashti incredulously.

"So they say," said her father.

"Lands sake!" said Temperance. "It seems like blood-money to me. Pore Len!"

As they all rose from the table, Mabella managed to slip away to her room, to spend the few moments before her tryst, alone. She looked out of her window and saw afar amid the boulders of Mullein meadow a form she knew, and the next moment she fled breathlessly from the front porch. A more sophisticated woman would have waited till the trysting time had come, but Mabella's heart was her helm in those days and she followed its guiding blindly, and it turned towards Lanty waiting there for her. *For her.* O! the intoxication of the thought! O! the gladness of the earth! the delight of feeling life pulsing through young veins!

And thus it was that as Lanty paced back and forth in patient impatience within a little space hedged in by great boulders, his heart suddenly thrilled within him as the needle trembles towards the unseen magnet; he looked up at the evening sky as one might look upon whom the spirit was descending, and then, turning instinctively, he saw a shy figure standing between two great boulders. He cast his hat to the ground and went towards her, bare browed, and, holding out his arms, uttered a sound of delight. Was it a prayer—a name, or a plea? And with a little happy, frightened cry of "Lansing, Lansing," Mabella fled to him. Nestling close to his throbbing heart, close indeed, as if she was fain to hide even from these tender eyes, which, dimmed with great joy, looked upon her so worshippingly. There are certain greetings and farewells which may not be writ out in words, and these untranslatable messages winged their way from heart to heart between these two.

The grey heaven bent above them as if in benediction. The stern outlines of the old boulders faded into the dusk which seemed to enwrap them as if eager to mitigate their severity. The soft greys of the barren landscape, the tender paleness of the sky, seemed to hold the two lovers in a mystic embrace, isolating them in the radiance of their own love, even as the circumstances of a United Destiny were to hedge these two for ever from the world. There were jagged stones hidden by the tender mists of twilight, and bitter herbs and thistles grew unseen about them, but to their eyes the barren reaches of Mullein meadow blossomed like a rose. Doubtless, they two, like all we mortals, would some day "fall upon the thorns of life and bleed," but together surely no terror would overcome them nor any despair make its home in their hearts, so long as across the chasms in the life-road they could touch each other's hands. The first rapture of their meeting vanished, as a bird soaring in the blue disappears from vision, which yet does not feel a sense of loss, because though the eye sees not the heart knows that afar in the empyrean the triumphant wings still beat.

"Mabella—my Mabella. You love me?"

"Oh, so much, so very much—and, Lanty, you like me?"

"Like! Oh, Mabella, since that day in the hayfield when I *knew*, you can't imagine what life has seemed to me since then—surely it is ages ago, and how I have thought of you! Dear, I can't say all I mean—but you know—Mabella, *you know*, don't you, sweetheart?"

"I hope so," she said sweetly, and then, with the inconsequence of women, her eyes filled with tears.

"Lanty—you—you will be good to me?"

"May God treat me as I treat you," said Lanty solemnly.

There was a pause, such a pause as when the sacramental wine dies upon the palate.

"I did not doubt you, Lanty."

"No, sweet one," he said; "I understand all about it. I will be good to you and take care of you, and, oh, my own dear girl, I am so happy."

"And I——"

And then lighter talk possessed them, and they recounted incidents, which, with the happy egotism of lovers, they chose to consider as important events because they had a special significance for them. The path to love is like a sea voyage. There are always more remarkable occurrences and extraordinary coincidences in one's own experience than in anyone else's, and these two were no exception to the rule. They discovered that upon several notable occasions they had been thinking *exactly* the same thing, and upon other occasions each had known *exactly* what the other was going to say before the words were uttered, and they talked on until they were environed in an atmosphere of wonder and awe, and looked upon each other startled by the recognition of their superiority, and the world was but a little place compared with the vastness of each other's eyes.

The dusk crept closer to them, the wings of night waved nearer and nearer, and Mabella, resting in Lanty's arms, sought his eyes for all light, and as they stood thus two other pairs of eyes watched them.

When Mabella had disappeared so promptly after supper, suspicion had stirred uneasily in Vashti's heart.

"Do you want me to help with the dishes?" she asked Temperance; "Mabella seems to have gotten herself out of sight."

"No," said Temperance, who was expecting Nathan, "I'll finish up that handful of dishes and everything else there is to do in half an hour."

Vashti betook herself to the garden expecting to find Mabella and Sidney there.

Both were gone.

Sidney, so soon as Vashti's personal influence was disturbed by the presence of others, fell again into a chaos of self-communings, and the devil which lurked there drove him forth into the wilderness; walking with the hopeless desire of escaping from himself he, ere long, found he was amid the barrenness of Mullein meadow. He wandered up and down amid its grotesque boulders till suddenly there came to him a sense of trespass.

"Put off thy shoes from off thy feet for the place whereon thou standest is holy ground," his intuitions whispered to him. He raised his eyes—looked out from his own heart and saw Lanty with Mabella within his arms, her eyes raised to his radiant with the ineffable trust of first love. Sidney stood spellbound, his heart aching within him. How sorely he envied the title-deeds to this enchanted country they had found, and possessed by divine right. Surely that meek man Moses endured sore agony as, foot-weary after long wandering, he looked upon the promised land, and looked only. It is indeed bitter to look at happiness through another man's eyes.

Sidney lingered some little time, till of their sacred talk one syllable came to him clearly; then he realized the sacrilege of listening, and departed; but surely the sky was very dark towards which he turned. Yet as he searched the sombre clouds before him the needle-like rays of a tiny star shone out environed by the darkness, and Sidney lighted a little lamp of hope at its beam.

When Vashti found the garden empty as last year's nests she never paused, but turning fled up to the little garret cupola whose windowed sides gave a view for a long distance in every direction, and hardly had she climbed to this eyrie before she saw two figures in Mullein meadow.

That was enough.

Vashti did not wait to study the picture in detail. Gathering her skirts in her hand she sped down the stairs through the garden and down the road like a whirlwind. Her thwarted will shook her whole being as a birch trembles in the breeze. Mabella had dared! When *she* had smiled upon him! As Vashti ran down the road she promised herself that she would give both Sidney and Mabella a lesson. Mabella would be presuming to Lanty next! So Vashti soliloquised within her angry soul as she climbed the stone fence of Mullein meadow and crept noiselessly towards where she expected to find Mabella and Sidney. She advanced stealthily, paying all heed to caution, and after duly ensconcing herself behind a boulder which she knew commanded a view of the little hollow she looked—and saw.... and controlled herself sufficiently not to scream aloud in rage; but vitriolic anger seethed within her heart, and for the time denied outlet, burned and cankered and tortured the breast which contained it. The first desire of her dominant nature was to fling herself before them in a wild accession of rage, and open upon them the floodgates of speech, but Vashti Lansing was not without a heritage of self-control. Long ago when her ancestress had been on trial for witchery, cruel persuasion had been used to make her speak—in vain. The torment of the modern Vashti was greater and keener, inasmuch as it came from within; alas! we are told, it is that which defileth; every proud drop of blood in Vashti's veins urged her to mocking speech; beneath the iron curb of her will she was mute, but the victory cost dear. So as Lilith, the snake-wife of Adam, may have lain in the shadows of Paradise watching the happiness of God-given Eve, Vashti Lansing stayed and watched sombrely, ominously, the joy of these two, and cursed them, vowing them evil, and promising the devil within her the glut of a full revenge—revenge for what?

Lanty had never given her cause to think he loved her, and Mabella had only veiled her love with shyness, not hidden it with guile—but—Vashti Lansing was supremely illogical. They had transgressed the unwritten statutes of her will. Did not that suffice to make them sinners above all others?—besides, like the poison which festers in the already wide wound, she realized in those moments of supreme mental activity that she loved Lanty, as women such as she love men, tigerishly, selfishly—Ah! they should suffer even as she suffered! She dropped her face in her hands, enduring the mortal agony of her baulked will, her misplaced, evil love, her bruised self-confidence, and shattering rage. And when she raised her head once more the scene had grown dark, the grassy stage whereon two mortals had lately mimed it in the guise of gods was empty, and she was alone.

She rose slowly to her feet wringing her hands in mute wrath. She looked around at the dreary field wherein she had endured such agony. Oh, that some yet more bitter blight than barrenness might fall upon it—some pest of noxious plants, some plague of poisonous serpents; oh, that she knew a curse potent enough to blast the grass upon which they had stood! But nature sanctifies herself; our curses are useless against her righteousness and rattle back upon our own heads like peas cast against a breast-plate of steel.

She entered the house calmly as was her wont. Within her heart was a Hades of rage; upon her brow the glamoured eyes of Sidney Martin saw the spectral gleam of the star of promise.

CHAPTER V

Sally, the small bondmaiden of Mrs. Didymus, stood at the garden gate of the parsonage.

No smoke curled up from the parsonage chimney, for the kitchen fire was out, Sally being much too occupied with other affairs to attend to her work that day. Work, in Sally's estimation, was the one superfluous thing in the world, and that she should be harassed with sweeping, and tormented with dish-washing, seemed to her an extraordinary and unjust dispensation. Sally had passed the first twelve years of her life in the slums, and her unregenerate soul yearned to return to the delights of dirt and idleness.

"Wouldn't I just love t' go back t' Blueberry Alley!" she said to Mrs. Didymus. "Wouldn't I just! My! I'd preach t' 'em!"

Mrs. Didymus' regret over Sally's first aspiration was quite lost sight of in her delight at the latter idea. She thought of "the little maid of Samaria," and smiled benignly upon Sally.

"That is well said," she answered; "some day, perhaps, you may carry the tidings. Little children have before now worked miracles. But over-confidence is a dangerous thing. You must not be too hasty, Sally; do you feel prepared?"

"Do I? Don't I jest? Sakes, I could tell 'em more about Hellfire and Damnation than ever they've heard of in all their born days. *Do I feel prepared?* Ruther! I'd just like old Lank Smith t' step up t' me, and begin aswearin', I'd let him hear a word or two that'd astonish him. He thinks he can swear!"

"Sally," said poor gentle Mrs. Didymus, hardly able to believe her ears, "Sally! Never let me hear you talk so again. The gospel is a Gospel of Peace."

"Gorspel o' Peace," said Sally, looking at Mrs. Didymus pityingly, "Gorspel o' Peace! Laws, mum, you are green! What chance d'ye think a Gorspel o' Peace 'ud have in Blueberry Alley? It's night sticks they needs there. Why, when I was a kid" (Sally had turned thirteen, but talked as if she was fifty) "there was missionars out o' count came to Blueberry Alley, but they mostly left a sight quicker than they came. There was a young priest came there, though, and the first day he went through the Alley the boys started t' have fun with him. Scrappin' Johnstone picked up a handful of dirt and hit him in the ear with it, and the priest got very pale, and he sez,

'It sez in the Scripter t' turn the other cheek t' the smiter,' and with that he turned hisself round, and Scrappin' Johnstone, thinking he had got a snap, let him have some soft mud on the other side. The whole Alley was on hand by that time. I was there. I mind I had a row myself a minnet after; but, anyhow, after Johnstone throwed the second handful he stood grinning in the priest's face, and the priest he got sickly white, and sez very quiet like, 'the Scripter sez t' turn the other cheek t' the smiter, and I've done that,' sez he, 'but,' sez he, 'it don't say nothin' as to what you're t' do after that,' and with that he pitched into Scrappin' Johnstone. He batted him over the head, and clipped him on the jaw, and biffed him back of the ear, and knocked him down, and stood him up and knocked him down again, then he laid him in the gutter, and stood over him, and told him he should behave hisself more gentle t' folks, and that fightin' was a sin, and that he shouldn't take advantage of strangers, and then he gave Johnstone and the Alley an invite t' come round and hear him preach in the chapel. The whole Alley's Catholic now. Gorspel o' Peace! That ain't the sort o' pursuasion Blueberry Alley needs."

Mrs. Didymus groaned in spirit, and held her peace, absolutely afraid of Sally's reminiscences. Sally and her ways were a terrible trial to the parsonage household, but good Mrs. Didymus could not contemplate the idea of permitting Sally to return to such an evil place as Blueberry Alley.

Sally was not well regarded in Dole, at least by the elect.

"One man can take a horse to the water, but twenty can't make him drink," was a saying frequently applied to Sally. This, being interpreted, meant that Mrs. Didymus could bring Sally to church, but that her authority, reinforced by the Dole frowns in the aggregate, could not make her behave herself whilst there.

"Sally," Mrs. Didymus would say, striving to temper severity with persuasive gentleness, "Sally! why do you behave so?"

"I dunno, mum," Sally would reply reflectively.

"But why don't you try to do better? Mrs. Ranger was terribly shocked by you to-day; she never took her eyes off our pew. What were you doing?"

"Nuthin'; she stared at me, so I stared at her, and now and then I'd cross my eyes at her for variety. Laws! I had the greatest mind in the world t' get up and turn round so's she could see my back. She seemed anxious t' look clean through me. Mrs. Ranger! Who's she I'd like t' know? I'd rather be a door-keeper in thy house, than eat fresh doughnuts with Mrs. Ranger," concluded Sally, piously loyal.

"Sally," said Mrs. Didymus, forgetting the main issue in the magnitude of the new offence, "that sounds terribly profane. I know you don't mean to be so, but don't use Scripture words like that."

"You're tired, mum, go and lie down, and I'll cover you up," said Sally, imperturbably.

"But, Sally, I'm very serious about this."

"Yes, I know, mum. Your head's real bad, ain't it? Lie down and I'll make you a cup of tea. Would you like a hot soapstone to your feet?"

Mrs. Didymus desired Sally's sanctification—she was offered hot soapstone for her feet.

Sally's assumption that rebuke sprang from illness was a very baffling thing to contend with, and Mrs. Didymus usually retired from the discussion beaten, to torment herself by wondering miserably if she was doing her duty by Sally.

If that worthy was not high in the estimation of the elders in Dole, she at least reigned supreme over the children. The bad ones she fought with and overcame, and the good ones she demoralized.

When Ted Ranger endeavoured to amuse himself by pulling Sally's tow-coloured hair, he received such a scratching that he never forgot it, nor did the village for some time to come, for he bore Sally's sign-manual upon his cheeks for weeks. When Mary Shinar's fifteen-year-old brother heard of this, and deigned to consider Sally a foe worthy of his prowess, the whole school gathered to watch the combat which ensued promptly when Jed Shinar called her a "Charity Orphan."

Sally precipitated herself upon him with such fury that he nearly fled from the first onslaught, and was extremely glad when the appearance of Mr. Didymus put a stop to the proceedings.

Jed's nose was bleeding, and mentally he was considerably flustered. Sally's hair was on end and her clothes were torn, but her self-possession was intact.

She retreated, led by the scandalized Mr. Didymus, but her fighting blood was up, and she called out opprobrious epithets to Jed till she was out of hearing—compliments which Jed's inherent and cultivated respect for the preacher forbade his returning in kind.

"He called me a Charity Orphing," she vouchsafed in explanation, when haled before Mrs. Didymus. "Now I know I'm a orphing, and I'm glad of it. Fathers and motherses mostly whacks the life out of you. But I won't have no freckle-faced kid calling me a '*Charity Orphing*!' Not if I'm well."

Mr. and Mrs. Didymus remembered the gruesome stories of demoniac possession, and breathed more freely when Sally left the room.

Upon the day of poor Len Simpson's funeral, Sally swung in luxurious idleness on the parsonage gate. Mrs. Didymus had gone early to the house of mourning.

Sally's tow-coloured hair, which was kept cropped to within five inches of her head, stood out like quills upon the fretful porcupine. Ever since Sally had seen a stray circus poster, with the picture of the beautiful Albino lady, with her outstanding locks, she had determined to arrange her own coiffure in like manner, upon the first favourable opportunity. So this morning she had rubbed her hair well with yellow soap, and combed it straight out, with a result which surpassed her anticipations.

About her waist there was a line of more or less white material. This marked the hiatus between her skirt and its bodice—a peculiarity of Sally's *ensemble.* When she stooped over, this white strip widened, giving one a horrible premonition that she was about to break asunder. When she stood erect, it frilled out around her like a misplaced ruff. Sally had bandied words amiably with every one who passed to the funeral, and when Sidney Martin almost stood still in his astonishment at her appearance, she was ready to greet him affably and volubly.

"Hello!" she said. "You're the Boston chap that prayed the rain down, aren't you?"

Sidney coloured quickly. The sting of his thoughts pressed home by the *gamine's* impertinent speech.

"Oh, don't be bashful," said Sally; "Mrs. Didymus says it was a powerful effort!" She uttered the last two words with impish precision.

"And who are you?" asked Sidney, feeling he must carry the war into the other camp.

"Me—well, you ain't been long in Dole, or you'd know me. I'm the maid of all works at the parson's." Then she harked back to the old theme. "So you really prayed in the church. My! You don't look as if you used bad words. Say, I thought there was some acters comin' t' the funeral? That's what I fixed myself up for. Say, how d'ye like my hair?" Sidney, despite his sad thoughts, could not forbear laughing as he replied:

"It's great, it's really great!"

"So I thought myself," said Sally complacently; then she added confidentially, "It's great for style, but 'taint much for comfort. I wonder when the acters 'll come. How d'ye s'pose they'll be dressed? When I was a kid in Blueberry Alley, I once went t' see Uncle Tom's Cabin. It was fine when Elizer went across the ice. My! it did jiggle. If you had been there, I s'pose you could have prayed it solid?"

An intolerable pang, absurdly disproportionate to its genesis, pierced Sidney's soul. His supra-sensitive nature was keyed to its highest pitch. The lightest touch upon the tense strings of his emotions nigh rent his being.

He turned swiftly away from the grotesque little figure, from the village street, from the house about which the vehicles were gathered thickly. An open road lay ready to his feet, and he took it, unconscious of its direction.

"There!" called Sally after him, "I've made you mad and I didn't mean to a bit. That's always the way with you religious people! You can't take a joke. It's maybe good for the soul, but it's mighty bad on the temper, religion is! And sakes! You mustn't mind me. I can't help being cheeky, "tis my nature to."" She finished with irate mockery, as the distance widened between them, and he did not reply. She was still looking after him, as he reached the abrupt bend in the road, and there he turned and bade her farewell in a gesture of unmistakable kindliness.

"Well!" said Sally, arresting her nonchalant swinging, with a jerk, "well, he ain't cross-grained, that's serting—Laws, I wish I had a civil tongue, but I hain't, so *Sigh no more, my honey*," with which she broke into a darkey song.

Sidney Martin went blindly along the path which chance had chosen for him, led by no other instinct save the old pathetic one, which prompts wounded creatures to crawl away to suffer unseen. Long ago, the human was equally sensitive, equally reticent; we are so no longer, but lay bare the plague spots on our soul with shameless candour.

But the nearer we are to God and Nature, the more prone we are to flee away into the bosom of the stillness, there to agonize alone; and not in vain do we put our trust in its tender sublimity. Nature never did betray the heart that loved her.

When Sidney paused, arrested more by an increasing sense of physical effort, which encroached upon his bitter self-communings, than by any conscious volition, he found himself upon a little wooded hill high above Dole.

Behind him stretched the whispering galleries of the wood, before him lay Dole, all its insignificance revealed.

The bird in the air is but a speck to our eyes; but how completely the position is reversed when from its airy altitude it deigns to stay its soaring wings and look beneath!

The greatest cities upon earth become but inconsequent masses when viewed from above. To Sidney's eyes, Dole looked scarce big enough to hold a heartache, yet how keenly its atoms felt!

And how little it disturbed the quiet heavens, the serene hills, all the suffering in the valley! This thought which, in one less in love with nature, might have unsealed fresh founts of bitterness, brought to Sidney's soul a beneficent sense of ultimate peace and strength. To him, one of Nature's own children, the mother tongue was very eloquent. And even in this hour of tense personal perplexity, he was able to gather some measure of consolation from the thought that in the end the jarring discordances of individual life would be absorbed into the grand symphonic song of Nature.

Nature is often impiously charged with unsympathetic indifference, by those who would wish to see all the heaven clouded over by their sorrow, a new deluge upon the earth because of their tears. But Sidney regarded his mother with reverent eyes, seeing in her seeming impassivity to his pain but a manifestation of the strenuous patience with which she waited to be renewed, looking towards that day when once again she would shine forth in all her pristine beauty, as she had been when first she was the bride of the sun.

"Scarred, and torn, and pierced, denied, disfigured and defaced by human hands, she yet smiles, and waits." So he said to himself. Truly Nature is justified of her children.

Flinging himself down upon the grass Sidney strove to find some gateway of escape from the awkwardness of his position, and gradually the accumulated nervousness of the last few days died away.

Nature's beautiful breast seemed to pulsate visibly and audibly beneath him, and he grew calm.

And so he lay for some time, and then slowly but imperatively other thoughts grew and gathered in his heart. The great primitive Want—spontaneous as it flamed up in the heart of the first man, resistless as its co-equal, Time, pinioned with the impulses of ages, sped by the impetus of æons—rose within him, knitting together all his strengths, all his weaknesses, into one desire.

He rose to his feet; surely his very bodily stature was greater?

He looked about upon the hills with brotherly eyes; deep in their bosoms beneath the grass the old elemental fires still slept. They could sympathize with him.

"Vashti—Vashti," he murmured. Out of his wildered musings there had grown the dream of the woman he loved, as the phœnix draws from out the ashes.

He looked again upon the village. Slowly, slowly winding along its ways, he saw a black stream of people and slow-stepping horses—Len Simpson's last journey through the familiar little streets. A chill shuddered through Sidney's veins. He had looked athwart the smiling champaigns of Love's country, and sullying its fairness he saw the black lake of mourners from which the sombre stream was flowing to the churchyard—saw it slowly gather there as the waters of a lake in a new basin. Here and there it had left stains along its course, as incurious or hurried units in the procession deflected towards their homes without waiting for the final solemnities.

It wrung Sidney's heart to think *she* was there in the gloom, whilst he, absorbed in selfish introspection, was aloof in the glory of the Sun. He must go down to her at once.

How little his generous soul dreamed that there was painful symbolism in that descent of his! That he poised upon the pinnacles, whilst she grovelled in the dust of her own desires, he never imagined. Indeed throughout all his life a merciful veil hung between these two, and hid the real Vashti from his loving eyes.

"Why didn't you come to the funeral?" asked Vashti, as he came upon them at the church gate.

"I went for an hour's quiet thought upon the hill," he said. "I had need of it."

"Wouldn't you like to see the grave?" she asked.

The latest grave was always "the" grave in Dole.

"Yes," he said, half dreamily. She led the way through the groups of men and women, who let the words die upon their lips as their glances followed the pair. There was little comment made, for Dole people were not prone to commit themselves, but they looked after Vashti and Sidney, and then into each other's eyes, and resumed their interrupted conversations—feeling all had been said which required to be said, when a young man and woman deliberately singled themselves out from the others. Vashti Lansing was most contemptuous of the trivial usages of the people among whom she had been born and bred; but she estimated very correctly their weight in the social system in which she had a place. And in this respect she showed wisdom.

She threaded her way swiftly among the graves, but in her abrupt avoidance of the mounds there was more indication of impatience at the obstacles presented than of tenderness towards the sleepers, whose coverlets, though heaped so high, could not keep them warm.

And presently they reached the corner, where, like a wan finger pointing reproachfully at the sky, shone the white obelisk above Martha Didymus' brown head.

The white shaft cast a slender shadow athwart a new-made grave at its side.

The red earth of the newly heaped grave was all but hidden with flowers, and a huge wreath had been hung upon the white stone; it had slipped down beneath the name of the dead girl, and hiding the rest of the inscription showed the one word "Martha" garlanded with flowers. Might one not dream that in the meadows of Elysium the young girl bedecked herself with fadeless flowers against the coming of her lover? Beside the two graves stood a group of clean-shaved, well-dressed men. Accustomed to mime in all guises, real grief found them awkward but sincere.

As Sidney and Vashti drew near they looked at the pair with interest. Vashti's striking personality had been singled out immediately from the throng of villagers at the funeral, but the eyes, accustomed to scan audiences, knew that Sidney had not been present.

"A friend of his?" asked a pale, handsome-faced man, with iron grey hair.

"No—but I have heard his story," said Sidney, in his soft, gentle voice.

"Well—he only asked for one thing—to be buried beside her," said the actor; then looking at the others he took off his hat, and in a voice, remembered yet for its melody in two continents, he repeated the matchless dirge,

"Fear no more the heat o' the sun,
Nor the furious winter rages."

Slowly, solemnly the beautiful words were uttered.

Their music mingled with the melody of his perfect voice, making them more than eloquent.

"Fear not slander, censure rash,
Thou hast finished joy and moan."

The words seemed almost personal in their application. The last word was voiced; slowly the little group turned away, following the man whose own life was clouded by so terrible a tragedy. Sidney stood bareheaded by Vashti, beside the two dead lovers, thinking that Len Simpson had been indeed honoured. To have Shakespeare's words syllabled above his grave is surely to the actor what the salute of the guns is to the soldier.

"Come," said Vashti softly. She was too politic to stay longer. No wise woman scandalizes the community in which she dwells. They advanced towards the others again, to find the tongues buzzing. There was a commotion amid the groups of women, which indicated that something out of the common order had occurred, which was indeed the case. For Mabella Lansing, unnoticed by the throng which was watching the actors openly and Vashti and Sidney furtively, had driven away with Lanty in his top buggy.

Here was daring with a vengeance!

Even Temperance Tribbey looked rather more grim than usual as she stood with Vashti waiting for the democrat to be brought round.

Fat little Mrs. Wither came gushing and bubbling up to Temperance with an affectation of confidential sympathy.

"My! I hope Mr. Lansing won't be long bringing the horses."

"Do you want a ride?" politely asked Miss Tribbey, as if oblivious of the fact that Mrs. Wither was that day driving her new buggy for the first time, and that her destination was diametrically opposed to the way the Lansings would take.

"Want a ride! Sakes, no," said Mrs. Wither, tossing her head. "But ain't you terrible anxious? I kin feel for you."

"Anxious about what?" asked Miss Tribbey coldly, eyeing Mrs. Wither steadily.

Mrs. Wither faded back into the crowd, giggling nervously.

"That Temperance Tribbey is the queerest woman!" she said to Mrs. Ranger as she passed.

Meanwhile, Vashti had been engaged upon the other hand by Mrs. Smilie, who was large, motherly looking, but dangerous. She had a way of enveloping her victims in a conversational embrace, and when she released them they were usually limp. Any information they had possessed prior to the meeting having been passed on to Mrs. Smilie.

But Vashti had refused the combat; having done so, however, with such a sorrowfully resigned expression that Mrs. Smilie felt her to be void of offence, and said afterwards:

"I was real sorry for Vashti Lansing. She was real humiliated. To think Mabella 'ud act up that way. Vashti looked really concerned; she's got a lot of sense, Vashti Lansing has! My heart jest ached fer her."

Mrs. Smilie's heart was always aching for somebody, but it did not tell much upon her general health.

As Nathan Peck, a sufficiently ridiculous figure in his suit of black diagonal, with the muffler superimposed, helped Temperance into the democrat, he squeezed her hand awkwardly, but avoided meeting her eyes; and she studiously looked over his head. Thus they acknowledged their mutual regret over Mabella's action.

Old Mr. Lansing was furious.

"Why couldn't you stay with your cousin?" he demanded of Vashti. "Going off buggy-riding from a funeral!! A fine speculation she's made of herself."

"I haven't seen Mabella since we left home," said Vashti softly, then she added deprecatingly:

"It's Mabella's way."

"Then it's a d——d bad way," said old Lansing, and then nearly choked with rage to think he had sworn in his Sunday black, which was so eloquent a reminder of his deaconship. He cut the fat bays across the haunches in a way that surprised them.

"Just wait till I see Lanty! And let her keep out of my sight!"

Sudden tears filled Vashti's eyes. She was sick at the heart with jealous pain. Sidney caught the glimmer of the tear, and felt a great throb of pity for this stately creature, who, fixed in her rectitude and dignity, could yet weep over thoughtless Mabella's little escapade. Needless to say Sidney saw nothing very dreadful in the two lovers driving home together; indeed, from the glimpse he had had of Lanty's face, he had no doubt but that after the burial of his friend, Lanty was in sore need of his sweetheart's consolation.

"Dear!" said Vashti, "I do hope Mabella will go straight home."

"I guess you hope more'n you expect, then," said her father irately.

Vashti sighed.

Miss Tribbey sniffed. The sniff expressed scorn, but it was wrongly applied by at least two of her hearers.

Miss Tribbey had no delusions about Vashti, and she knew the girl was doing all she could to irritate her father against her cousin.

"M'bella's young and foolish," said Temperance grimly, but with apologetic intent in her voice.

Vashti gave her a venomous side glance and sighed again.

"It's the French grandmother coming out in her. Gee! It takes ages to kill a taint, and then every now and then it crops out," said old Lansing.

"Yes," said Vashti, "that's what Mrs. Smilie said. 'It's the French in her,' she said." The moment Vashti uttered this she bit her lips angrily, for a swift change passed over her father's face, and she knew she had made a mistake.

"She did, did she?" roared old Lansing, purpling with rage. "She did? The idea of these mongrel Smilies setting up their tongues about the Lansings. Lord! I mind well her father drove about the country collecting ashes for a soap factory. She ain't fit to black Mabella's shoes—that woman. What did she do when she quarrelled with Mrs. Parr? Went and threw kittens down her well, and they most all died before they found out 'twas the water. She'll talk about the Lansings, will she——"

Old Lansing rarely began to gossip, but, when once fairly started, the revelations he made were rather startling. He continued until they reached home.

Lanty and Mabella walked side by side up and down the wide sandy path from the front door to the garden gate. A look of deep and grave happiness shone upon their faces; both were looking at their future from the same standpoint. There was a hint of timorousness upon the girl's face, an occasional tremor of her sweet mouth, which told that all terrors were not banished from the Unknown, into whose realms the man at her side was to lead her; but hallowing her face there was that divine trust which transfigured the Maid Mary into the Madonna.

"I am going to speak to uncle now," said Lanty, "and if he is pleased we will go for a drive after supper to-night."

"Yes," she said; then "Lanty." He looked at her; she uttered no other word; her eyes slowly filled with tears.

"Mabella, you trust me?"

"Absolutely," she said, and the tears, brimmed over by a tender smile, glistened upon her cheeks.

"My angel," he said, and gave her a look of adoration, then turned away, and went striding round to the side of the house where the others were alighting from the democrat waggon. Old Lansing looked up sharply as Lanty drew near. Something in the young man's face held him silent an instant.

"I'm coming round to the barn with you," said Lanty; "I want to speak to you."

Sidney turning away heard the last words. He could not forbear a look of sympathetic comprehension. Lanty flushed to the eyes, and from that moment was a staunch and faithful ally to Sidney.

"She's up on the landing," said Temperance, as, a few minutes after, Lanty, pale and eager, entered the kitchen. Lanty had not spoken—nor did he now, but he went up to Temperance, put his hand upon her shoulder, and gave her a hearty kiss. Then he turned and went up the back stairs three at a time. Through the back hall to the great dusky silent landing, and there a little figure waited trembling.

"What——" she began, and then her quivering lips were silent.

"It's all right," said Lanty, in a voice he hardly recognised as his own. "You are mine—mine."

She laid her face against his breast and there was silence between them. And whilst they supped of Beatitude, proud Vashti Lansing, pale as old ivory, was walking up and down the path their happy feet had trodden such a short while before, tasting the very bitterness of Marah, but compelling her proud lips to tell Sidney Martin the story of their French ancestress.

Vashti Lansing had more than one heritage from the murdered witch wife. The courage which had kept the old Vashti calm and contemptuous before the fagots, upheld the modern Vashti in her time of torture. It is the fashion to sneer at grandfathers—among those who have none. Nevertheless, the fact remains that there are very few Esaus, although there are always plenty of Jacobs, ready to buy birth-rights if money will do it.

It is a good sign if a family guards its traditions carefully. The types presented in these oral picture galleries are sometimes not the best types, but they at least shine forth distinctly from their background, and be their light clear or lurid it is by these beacons that we are guided back to the beginnings of character. How much more eloquent and rich a language is in its meaning to us when we know its root words! that we are guided back to the beginnings of character when we can comprehend its genesis, and trace the subtle transmutation of one characteristic into another; the change of physical courage into moral strength, or perchance—the retrogression of simple tastes to penuriousness, or the substitution of intellectual enthusiasm for the fires of ardent passions. Family tradition is the alphabet of all history! What contrasts are presented amid the pictures thus preserved! And surely there was never greater difference between two ancestors of one house than existed between old Abel Lansing, the donor of the Lansing legacy, and beautiful Germaine Lansing, the wife of pious Jason Lansing. Jason Lansing had wooed and won and wedded his wife whilst he was in England doing the errands of the little colony of wanderers beyond the sea. How his choice fell upon frivolous Germaine, why she accepted her grim lover, none can guess; but certain it is they were an ill-matched pair. Our sympathies are inclined towards the gay little Frenchwoman who sang her chansons of love and ladies' lattices in the very ears of the elders, and rustled her brocades beneath the disdainful noses of their winsey-clad wives; but the community in which she lived regarded her advent in their midst as a "dispensation" of a peculiar and trying type. Jason Lansing could only sustain his good opinion of himself by remembering that even the patriarchs had not displayed entire good judgment in the bestowal of their affections. Her memory still survived among the Lansings—a frail ghost hung with scornful garlands of forbidden frivolity, and when any of the name outraged the traditional proprieties, it was said that the cloven hoof of French levity was showing itself once more. And with such tales as these, Vashti Lansing beguiled the dewy twilight hours for Sidney Martin, and stole his heart away, whilst her own burned and yearned for a love denied it.

CHAPTER VI

After the day when, alone upon the hillside, Sidney watched Len Simpson's funeral wind along the narrow ways of Dole, there ensued for him a sweet calm interlude—a tranquil period, yet surcharged with potentialities.

It was the space between the casting of the grain into the ground and the first blade. At such a time there is no stir upon the surface of the earth, yet in its brown bosom the vital germinal growths are beginning; the husk of individuality is bursting, the tap-root of deeper sympathy is searching for sustenance; and at last upon some happy morning a green glow gladdens the sky, and we say: "Lo, the new grain!" and offer thanks for the promise of the gracious harvest.

But all the after-vigour of the plant depends upon that silent time in the darkness. So the whole fabric of Sidney's after-life was built up from the beginnings made in that uneventful month, whose days are difficult to chronicle, as beads which slip adown the string and mingle with each other are to count. It passed like a lover's dream to Sidney, to be remembered afterwards as a season of peace and happiness whose source and sense eluded analysis.

The calm happiness which encompassed the lives of Mabella and Lanty lay like a benediction upon the house, and the hearts beneath its benison rested for the moment like a congregation hushed after the last Amen, and not yet surrendered again to the worldly cares and sordid joys which wait without the sanctuary doors.

But as one of the peaceful congregation may writhe in the hair shirt of personal perplexity, so Vashti Lansing beneath her calm smile suffered agonies in those days.

Is there any torture more poignant than the cry of "Peace, Peace" when there is no Peace?

She was very pale, the insolently perfect oval of her face had fined a little, there was a hint of a break in the suave curve of her cheek, and this, albeit an imperfection, lent her beauty a new and subtle charm of appeal.

She was very quiet, too, and now and then a tender wistfulness dimmed her eyes, softening the majesty of her brow alluringly. When Sidney saw this he felt his heart go out to her more strongly than ever.

"Unconsciously," he said to himself, "her sweet, strong nature covets the joy of loving and being loved"; and there welled up within him that indulgent and protective tenderness which all good men feel for the women they love.

Vashti Lansing had never appeared so gentle, so womanly, so good, as at this juncture when all the evil within her was rising, and gathering, and forming into malevolent purposes. Some deadly creatures take to themselves the semblance of flowers that they may sting their victims unaware.

Mabella and Lanty were together continually. It was very pretty to see her shy eagerness for his coming, his open happiness at her presence. Temperance was always busy with her housework, to which was added now the cutting and hemming of Mabella's household linen. For Temperance had long saved egg-money and butter-money for such an emergency, and, delighted at the prospective union of her two favourites, she fell to the work eagerly. Mabella tried to help, but her usually busy fingers were rather idle during those first halcyon days. She let her hands fall in her lap with the needle between her fingers, and slipped away into a dream, leaving all earthly considerations far behind. If a word or a smile reminded her that mortals were peeping into her paradise, she would rise and steal away to the little shadowy room, from the windows of which she had seen him waiting in Mullein meadow, and there, chiding herself for over great delight, she would strive to bring down her great joy to the basis of every-day fact. "We love each other," she would say, stating the fact in bold terms, "we love each other," and by the time she had said it twice her face would shine again with the glory of the thought, and the words ceased to become words to her, and became only the sighing of Love's mouth. What a simple figure Mabella Lansing presents upon the little stage whereon these people trod, beside the splendid and forceful personality of her cousin Vashti! What an ordinary and commonplace product of ordinary and commonplace conditions Lanty Lansing seems beside Sidney Martin, supersensitive, morbidly idealistic, a Sir Galahad, bearing the white flower of a stainless life and giving it into the hands of a wicked woman to work her will upon it!

Yet though the love of Mabella and Lanty was but "the homespun dream of simple folk," still the very gladness of it makes it precious in this world, where even the divine passion has grown a little hum-drum, and where the ashes lie whitely upon the divine fires.

But perhaps the world will shake off its lethargy when the new century begins, and even now there may be smiling in his cradle the Shakespeare whose breath shall blow the embers again into flame. Surely it is simple, natural kindly souls like Mabella and Lanty who perpetuate fidelity, honour and trust upon the earth; and eager, pure, unselfish souls like Sidney Martin who transmit the glorious impetus of aspiration from one generation to another.

It is hands like theirs which crown the years with enduring chaplets, and brush from the brow of the aging century the dishonouring garland of senile sins which are like toadstools, the efflorescence of decay.

Old Mr. Lansing having become better acquainted with Sidney, had ceased to regard him as "company," and had relapsed comfortably into his own ways. Reading his weekly paper, gossiping with Nathan Peck (who, being the village carpenter, always knew the latest news), and going to bed when the grey died out of the twilight sky.

Vashti and Sidney were thus left much to themselves.

The "odd" horse having effectually lamed herself by stepping on a nail, driving her was out of the question. To break a team upon any frivolous pretext would have been a scandal in Dole, so Vashti and Sidney were kept busy going errands. They went to the post office twice a week; they made pilgrimages out to the far-away hill pastures, where the young cattle grazed, to count them, and report upon the depth of water in the little brown pool where they drank.

What glorious days these were to Sidney; what rapture to stand upon some little eminence with the wind, "austere and pure," blowing across the valley upon their faces; with Vashti beside him, her eyes meeting his with sweet serenity, or looking vaguely forth far across the country, as if to seek out some haven remote from lesser mortals. So Sidney translated her thoughts, but in the original there was writ only bitter speculation as to whether *they* were together—if his arm embraced her, if their lips—Ah! it was of no remote haven that woman dreamed.

They gathered great fragrant bunches of spearmint and tansy, smartweed and pennyroyal for Temperance, searching for the scented herbs as children search for joy; and as the memory of childish pleasure lingers long with us, so the perfume of the aromatic herbs clung about Vashti's garments and Sidney's sleeves. Never again could Sidney know the wholesome odour of any of these plants without seeing Vashti, her tall figure in its faded blue gown standing straight and strong against the sunlight, with a huge bunch of greyish-green clasped to her breast, above which her face, fit for Burne-Jones' most mystic, most beautiful maiden, shone out palely. About her was no mystery of birth or circumstance, no halo of romantic environment, but her whole personality was eloquent of mystery, the sphinx-like riddle of sex presented in a new and strongly individualized type.

Their many expeditions together begot a sense of companionship which was inexpressibly precious to Sidney. True, as he realized, it sprang rather from circumstances than from the manifestation of any personal predilection upon Vashti's part; and yet, humble as he was before the woman he loved so blindly, he could not but be aware that she brightened perceptibly at his approach, and was always very willing to undertake any message or errand with him.

So she fooled him exquisitely, solacing her wounded pride thus. Whilst he, too great-hearted to pry for petty faults, dowered her lavishly from the generosity of his own noble nature, with all the classic virtues.

With what reverent fingers we hang virtues upon the lay figures of our imagination! How we becrown them, and worship them and offer them the incense of our efforts! Yet, it is pleasant pastime, and sanctifying too, for incense purely offered hallows the hand which gives it, perchance more than the God to whom its smokes ascend.

All this is well, and though the world gape and wonder at our adorations, what is that to the devotee? Only, to some of us comes the hour when with trembling hands we must undrape our false gods, lay bare their feet of clay to jeering eyes, fold away the rich draperies in which our love has clothed them as a mother folds and hides away the garments her dead child wore, and carry the manikins to the grave.

Happy for us if we can bury our dead decently; but bury them never so deep, they rise and walk down the vistas of our happiest hours, infecting their sunshine with the pollution of dead faith.

During these long walks together Vashti and Sidney talked much, and of more vital subjects than are generally discussed between young men and women. The fashionable chit-chat about theatres and plays, receptions and fashions was utterly missed from their calendar of subjects.

Now and then, Sidney, being a man, could not forbear to let her know how beautiful he found her; but empty compliment, the clipped coin of conversational commerce, he did not offer her; nothing but pure gold minted by her sweet looks in his heart was worthy of her acceptance. Thus they fell back upon the old immortal themes which have been discussed since the world began. They looked at life from widely different standpoints, but their conclusions were equally forceful.

Vashti Lansing had nothing of the simpering school-girl about her, and none of the fear which makes women reticent sometimes when speech would be golden.

It has been said that to know the Bible and Shakespeare is to have a good English vocabulary. Vashti did not know Shakespeare, but she knew her Bible thoroughly. Her speech, unweakened by the modern catch-words which, if expressive, are yet extraneous and dangerous growths, had all the trenchant force of the old Anglo-Saxon, with much in it too of imagery and beauty; for she did not fear to use such metaphors as nature or life suggested. Steeped in the stern Mosaic law, she knew well the stately periods of its prophets. The gentle Christ-creed of forgiveness did not find favour in her sight. "An eye for an eye, a tooth for a tooth" was a judgment which she said only timorous souls feared. She read with grim delight the tales of the kings, with their feet upon their captives' necks; an evil sympathy with their triumph lighted her eyes with wicked light. What a spouse she would have been for one of these cruel kings! she thought sometimes. And she applied a relentless utilitarian philosophy to life. The weakest go to the wall and the strong triumph. She accepted that with the stoicism which springs from conscious strength, but in her system she rather confused strength with righteousness. She watched the movings of life about her with cold, curious eyes, and yet her philosophy of life was but an expanded egotism. She comprehended only those sets of actions which might have taken place had she given free rein to her own inclinations; she judged of all motives by the repressed impulses of her own bosom. She scrutinized others unsparingly, prying into the most sacred griefs, the most holy joy without shame or remorse, and she did not spare herself more than others.

The dim, terrifying impulses and visions which girls put behind them, shudderingly and uncomprehendingly, hiding them away with the other spectres which people the realm of the unknown, until such time as life's meanings shall be expounded in a sacred mystery play of sense and spirit, she marshalled forth into the light of day and considered calmly and cynically.

She applied the foot-rule of her own lymphatic temperament to the morals of her fellows and was never disappointed when they fell short. She was well versed in all the wisdom of the Pharisees, and at the sewing circle talked always to the older women, and was never found in the corner where the clear-eyed girls whispered together.

And quickening and vitalizing all her existence there was that sense of Power. Power uncomprehended, undeveloped, yet there; and as a thunder-cloud gives premonition of its potent force even before the brand leaps from its cloudy sheath, so Vashti Lansing's personality was instinct with potentiality.

This was the woman Sidney Martin, idealist and dreamer, loved.

The days sped swiftly, the present lapsing into the past, the future flying forward with the unique tirelessness of time.

How wrong to typify Time with hoary head and tottering limbs. Crowned with the vigour of eternal youth, does he not leap forward triumphantly like the messenger of the gods fresh plumed with flame? Ah, he is not old, but young and swift. Strive if you will to stay his flight for but one single precious instant, stretch forth your hand while yet his wings brush your face, and ere the fingers may close upon his pinions, he is gone, leaving but the *largesse* of lost days.

The harvest was done, the ploughshare and the harrow were tossing the earthy bed for the new grain. Day after day, through the clear air, there came from different points the blowing of the traction engine which dragged the one threshing mill in the section from farm to farm.

It was the custom of the neighbourhood that the farmers should assist each other with the threshing. Sidney was charmed when he heard this—how idyllic it was this community of helpful effort! To be strictly truthful, this custom had its genesis in less worthy reasons than he imagined, the simple fact being that in the little hide-bound community there were no odd men left unemployed, therefore as labour could not be hired the farmers perforce clubbed their efforts.

"I say, girls," said Lanty, rushing out from his uncle's big barn to where the two girls and Sidney stood beside the engine, "I say, isn't that engine exactly like Mrs. Ranger in church?" His face was begrimed with dust, thistle-down rested whitely upon his yellow hair, his blue eyes were alight with hope and happiness and that exaltation which a strong man feels in effort. The girls shook their heads warningly, but laughed.

The traction engine, its wheels shackled, puffed and panted with a ludicrous simulation of bottled-up energy, and to the minds of the three young people it was decidedly suggestive of the irate patience expressed in Mrs. Ranger's attitude when placed in conditions where she could not answer back.

Nathan Peck, watching the engine, stored up the saying for Temperance's delectation, and wished she had come out with the girls.

Above the rattle and hum of the threshing mill sounded the hoarse voices of the men shouting jokes at each other—threshing time being always a jovial season. A good or bad harvest meant often life or death to these people; but, having done their best, they could but accept the results. It was a point of honour to accept unflinchingly the verdict of a poor yield, yet many wives could tell of despairing hours when, after their neighbours had departed, husband and wife essayed to reconcile ways and means.

Clouds of golden dust, starred here and there by a silver thistle-down, shimmered out of the barn door; there was an aroma of crushed straw, a scent of charred wood from the engine fire, a sense of eager, healthy life.

The swallows flew agitatedly above the barn, yearning over their clay nests beneath its eaves.

"What are you doing?" asked Vashti.

"Measuring," said Lanty. "Uncle said he'd take the bushel for a little though when he saw your petticoats out here——"

"Who's in the mow?"

"Ab Ranger is cutting bands, and he's let my bone-handled pruning knife go through the mill; Tom Shinar is feeding; there's three on the mow and four on the stack."

"How is it turning out?"

"Splendidly, no straw to speak of, but finely headed—like you, Mabella," he whispered, blushing through the dust.

"Come on here, Lanty," roared a voice from the barn. "You can spark in the noon-spell if you want to."

A laugh followed. Mabella blushed hotly, and as a maiden is expected to do under the circumstances, looked absently into vacancy.

"Well, you'll be too busy eating in the noon-spell to notice," Lanty called back to the unseen speaker. This, being the retort courteous, was received with applause.

"Well, I must go, girls; uncle's back will be aching by this time toteing that bushel. I hope you've made heaps of good things for dinner, we're all hungry as hunters."

"Trust Temperance for that," said Sidney.

"Yes, indeed," said Lanty. "Ta-ta, girls."

"Lanty," said Mabella, "be careful of the belt."

"Surely," he said, his voice softening. The next moment his strong, lithe figure had swung jauntily through the narrow space between the broad whirling belt and the door.

"Nathan," said Mabella, "Temperance wants you to get someone to mind the engine for ten minutes before dinner, so that you can come round and carve the meat."

"I'll be there," said Nathan, then he added with an irrepressible and comical self-importance:

"Meat ain't worth puttin' teeth into if it ain't cut up proper."

"That's very true," said Sidney, who felt a great kindliness in his heart for this patient lover.

"Well," said Mabella briskly, "I'm going round to help set the table." Having seen Lanty, Mabella wished to get off alone to think over his perfections, which impressed her afresh each time she saw him.

"O! can't you come for a little wander?" asked Sidney of Vashti. "There's nothing to be done in the house; besides, that imp from the preacher's is there, and I'm sure she is a host in herself."

"Yes," said Vashti, her voice more than usually vibrant. "Yes, I will come."

She was very pale. She turned away as Jephthah's daughter turned from the promise of her bridal bower. For, during these few minutes of idle speech amid the whirr of the threshing mill, Vashti Lansing had taken her final decision. She would marry Sidney Martin; but on her own terms, she added to herself. And then she went with him across the stubble, where the late rains had made a phantom spring of fresh green grass and over-eager weeds, which were putting forth their tender tops only to be a prey to the first sneering frost.

Ah, how futile and inconsequent it is to trace laboriously the windings of cause and effect; a touch often sends one over the precipice, and a smile, a sigh or a silence brings us face to face with Fate. Can we by searching find these things?

And Sidney, too, felt the fateful words trembling upon his lips, a keen envy of personal happiness possessed this man, who so rarely sought his own good. A great longing to stand as Lanty had stood, with the promise of life's fulfilment at his side.

Sidney and the woman beside him walked across the stubble to where a little belt of scrubby oaks followed the course of a ditch between two fields; here and there a vivid red patch against the underwood showed a dogwood bush. Here and there an elm tree sprang up spire-like above the lower oaks.

"See," said Sidney, "that row of elm trees. Can you not fancy that upon just some such day as this the seed was sown? Does it not give a delightful sense of the continuity and endurance of nature's miracles to think that a gentle wind, such as now stirs their topmost leaves, chased the seed vessel playfully along the ground? The wind laughed then, thinking it was making fine sport of its little playfellow, but see, at every pause a seed was dropped, and like an egotistical king who marks the stages of his journey, the fragile cluster of seed has left its memento. You have seen the seed of the elm tree?"

"Yes, it resembles a hop. I suppose the seeds are between the little scales. I can fancy it fluttering along the ground like yon leaf."

"Yes," he said, delightedly, and then, pleased with her comprehension of his thought, he looked far across the field. After all Mabella had not been in such a hurry to get to the house. She was running up and down like a child with the little brown calves in their special paddock near the house. Her sunbonnet was in her hand, her hair glittered in the sun like ripe wheat. From her Sidney's eyes turned to Vashti, and his very heart stood still, for dimming the splendour of her eyes two great tears hung between her eyelids. There was no quiver of lip or cheek, no tremour of suppressed sobs; her bosom seemed frozen, so statuesque was her pose.

"Vashti!" he said. It was the first time he had called her by name—used thus the one word was eloquent.

"Don't!" she said. "I—will—come—back to the house presently."

Sidney, his heart wrung, took his dismissal without further speech. He went a few steps from her, then turning went swiftly back.

Her tense attitude had relaxed. She was leaning against the grey bars of the fence, a crimsoned bramble twining round one of the upright supports hung above her as a vivid garland.

"Vashti!" he cried, "I can't leave you like this."

"Not if I wish it?" she asked, and gave him a fleeting smile, beautiful as the opalescent glimmer of the sun through rain.

It shook the man to his soul. He stood for a moment blinded by the glamour of her beauty, then left her again. This time he did not look behind, but strode triumphantly across the fields, for he felt that smile had given him definite hope.

Sidney, despite his perfections, was only man. For a moment he had forgotten her tears; then remembering, he said to himself that soon he would kiss away all tears from her eyes.

The best of men are prone to consider their kisses a panacea for all woman's ills. Perhaps, with the irrefutable logic of the homœopathists, they argue that what produces an ill will cure it!

"Tears, idle tears, I know not what they mean;
Tears from the depth of some divine despair
Rite in the heart and gather to the eyes
In looking on the happy autumn fields,
And thinking of the days that are no more."

The lines sprang spontaneously to his lips. This was the secret of Vashti's tears. How often he had felt that almost intolerable regret, begotten by the recognition of the evanescence of beauty. And Vashti with her splendid natural soul must feel with treble keenness all these things.

Doubtless to her the crimsoning of the leaves was as the hectic flush upon an ailing child's cheek to mother eyes. "*The days that are no more*," ah! could it be she thought of the days when the grain was growing high, the first days of their companionship? Deluding himself thus with futile fancies he turned slowly, slowly towards the house, arriving to find Vashti already there in the midst of the housewifely bustle.

Whilst the visionary Vashti bore him company, the real Vashti had passed him unseen. So it was ever. The real Vashti eluded his vision; her place was filled by a mimic Vashti created of an ideal and his love, and tricked out in all the virtues.

At the house every one was busy. The preparations for dinner were approaching a crisis. Temperance, with a look of ineffable importance such as only a managing and forehanded woman can wear upon such an occasion, was cutting pies, piling plates with biscuits, arranging pickles in glass dishes, and between whiles taking flights to the oven, where a huge roast was browning.

Mabella was arranging the table with knives and forks; she reckoned up, six or eight times, the number of people to set for, substracted two for the ends, and divided to find how many for each side. Mabella had no head for figures, so she made a mistake in this process; but as the basis of her calculation was wrong the result was correct. An unexpected thing! But Mabella, cheerfully confident in her methods, had no thought of all this; she trotted about the table with the gladness of one who does not save steps.

Vashti was bringing chairs out from the other rooms to complement the number in the kitchen; and Sally, the preacher's handmaiden, was arranging the tin basins with soap and water for the men to wash in, and varying the monotony by tantalizing the chained-up mastiff till he was nearly crazed to get at her, drawing back to his kennel door and launching himself forward with magnificent disregard of the chain which at each attempt jerked him off his feet.

Sidney leaned against the door-jamb watching the homely scene with just the faintest tinge of proud proprietorship in his eyes when they rested upon Vashti.

Presently she came and stood before him. Her figure was so suavely graceful that her most ordinary movements took on an artistic significance. Just now her attitude was that of a queen who fain would ape the serving maid, but who could not cast aside her sovereignty.

"Will you sit down with the men?" she asked.

"Your father does, doesn't he?"

"Indeed, yes."

"Then I will also."

"Then I'll wait on you," she said, and primmed her mouth into a quasi-humble expression.

"If you do——" his grey eyes dilated.

"Yes."

Just then Nathan came round from the barn.

"They'll be here in ten minutes," said Vashti, and hurried away.

Temperance, flushed with housewifely pride, had the big carving platter ready with the steel beside it. The latter was a concession to appearances, for Temperance always sharpened the knife for Nathan in a peculiar fashion of her own. When Nathan entered she was sharpening it vigorously on the back of the kitchen stove.

"Well," said Nathan, "here I be; where's the water?" He had seen the basins upon the apple-tree blocks, where they had stood for time out of mind at the Lansing threshings, but he thought Temperance might be prompted to come and get it for him.

Temperance paused in the sharpening process, but at that moment a tow-head appeared at the door.

"Here 'tis, Mr. Peck," said Sally, "right here under the shade; fresh water, sweet water, well water. Come up, run up, tumble up, anyway t' get up; here's where you gits water. Step up, ladies and gents. Everything inside as represented on the banners, and all without money and without price," concluded Sally, putting a frosting of the parsonage piety upon the vernacular of the Blueberry Alley dime shows. Mabella, Vashti and Sidney laughed. Temperance resumed her knife-sharpening with a click.

"That child will come to no good end," she said to Nathan, when he re-entered.

"She won't," agreed Nathan with some asperity; his waistcoat and shirt were drenched. He had asked Sally rashly to pour a dipper of water on his head to "rense him off." Sally complied with alacrity, only she emptied a pailful over his bent head instead of a dipperful.

"Drat that young 'un," said Temperance, enraged at this. "I believe, I really do, that Mrs. Didymus sent her over here to be shet of her for a day, and if this is a sample of her doin's I don't know as I blame Mrs. Didymus, but if there's any more goin's on I'll trapse her back quicker."

By this time the roast was out of the oven, and Nathan began his work with the enthusiasm of an artist.

Nathan was always greatly in demand when there was any carving to be done, and he was very proud in a candid childish way of his proficiency. Perhaps his practice with the plane and the drawknife stood him in good stead, for certainly Temperance was justified in thinking proudly that no man could carve like her Nat.

"They've blew," announced Sally, tumbling into the kitchen in great excitement. This was somewhat unnecessary information as the whistle was making itself perfectly audible; ere its shrill echo died away the men, begrimed and laughing, came round the corner of the barn and were soon spluttering in the basins.

Lanty came into the back kitchen, but the voice of one of the men brought him out of his retreat, and in five minutes they were all at table.

Old Lansing at one end with Sidney at one side. Lanty at the other end with Nathan beside him.

"Open the ball, Nat," said Lanty, passing Nathan the platter. Nathan helped himself with the deprecating modesty of one compelled to pronounce judgment upon his own handiwork; then the platter made the round of the table in pursuit of the one which had started from Mr. Lansing's end.

"Guess you had something to do with this, Nat," said Ab Ranger. "I know your shavings."

Nathan admitted the impeachment.

"Well," said Sidney, "we can't beat that in Boston."

And Nathan ate vigorously to hide his embarrassment. The girls flitted about seeing everyone was supplied. Did calm-eyed Vashti know what she did, when she bent over between Sidney and her father ostensibly to remove an empty plate, and let her palm rest as if by chance for a moment on Sidney's shoulder? Did ever electricity shoot and tingle through the veins like that touch? He watched her as she passed serenely along the other side of the table, and longed for the moment when he might have speech with her.

Temperance poured the tea and coffee in the back kitchen. Sally performed prodigies in carrying it to the table, and grimacing, as she set it down, behind the unconscious backs of the recipients.

Sidney won golden opinions at this dinner by his frank friendliness.

"He ain't big feelin', that's one thing," the men said to one another as they swaggered out to rest the noon-spell under the trees.

Lanty and Sidney with great affectation of helpfulness asked the girls to stand aside and watch them clear the table. Temperance was not to be seen, they would surprise her when she arrived. They succeeded beyond their expectations.

"It isn't such a job to clear a table as you'd think," said Sidney complacently to Lanty.

"No, 'tain't for a fact. I've seen girls take half an hour at it."

The two young men had cleared the table by removing the dishes and *débris* indiscriminately and depositing them upon the table in the back kitchen.

When Temperance returned from a little chat with Nathan beside the smoke house, she eyed the chaos upon the table wrathfully.

"Laws!" she said. "Of all the messes! Lanty Lansing, ain't you ashamed to be so redecklus? And them girls standin' gawkin' and laughin'! As for you," eyeing Sidney severely, "I should ha' thought you'd more sense, but blessed is them that has no expectations! Lanty! Are you or are you not feedin' that brute with good roast? Where's the cold meat fer supper to come from, I'd like to know?"

No one volunteered a response till suddenly Sally piped forth in her thin reedy voice:

"Take no heed for the morrow what ye shall eat, or——"

"You blasphemous brat!" said Temperance, her wrath diverted to another channel.

Sally subsided into silent contemplation of the dish of pickled beets from which she was helping herself with pink-stained fingers. Temperance was not Mrs. Didymus, and Sally in many combats in Blueberry Alley had learned to gauge her antagonists.

The offended Miss Tribbey left the back kitchen in indignant silence and set about arranging the table for her own and the girls' dinner, murmuring to herself meanwhile a monologue of which such words as "messes," "sinful," "waste," and "want o' sense," were distinctly audible.

"I don't believe that was really an unqualified success," said Sidney to Lanty.

"No," said Lanty, "I don't believe it was. What did you mix everything up for?"

"How did I know they were to be separated? What did you feed the dog with the roast for?"

"Did you ever see such an imp as that Sally?"

"Never," said Sidney. "But Temperance squelched her!"

"She did," said Lanty. "I say, wasn't she ripping?"

Meanwhile Temperance's short-lived wrath had died away, and she was pressing food upon Sally in quantities calculated to appal any but a Blueberry Alley child.

Temperance rose in the midst of her second cup of tea, and, going up stairs, came down with a large fresh bandana handkerchief. She went out to where Lanty and Sidney stood talking.

"Here's the handkerchief you wanted to keep the dust out of your back," she said with ill-assumed hauteur. Lanty took it with laughing penitence on his face.

"I say, Aunty," he said, "would you ask Mabella to put it on?"

Miss Tribbey's severity relaxed; a vain-glorious satisfaction stole over her face in a smirk. To have Lanty call her Aunty!

Certainly Lanty Lansing "had a way" with women that was well-nigh irresistible.

"Yes," she said, then with comical apology, she addressed herself to Sidney. "Them children is a most tormented trouble, 'specially when they meddle with things they don't know nothing about."

"That's so," agreed Sidney with emphasis, and Temperance, highly delighted with her Parthian shot at him, departed.

And presently Mabella came to the door, a *riante* little figure, and demanded with mutinous affectation of indifference:

"Did any one want me?"

"Yes, badly," said Sidney, and took himself off to the garden, laughing.

"That's true," said Lanty. "I did want you badly."

Her eyes were wavering beneath his masterful regard, but she said—"Oh, you *did* want me! Don't you now?" The words were brave, but her eyes fell.

"Mabella," he said—silence. "Mabella, look at me." Slowly she raised her eyes and crimsoned. "Do you know now?" he asked lovingly. "Ah, what a wicked teasing bird it is when its wings are free, but after all they are gone to the barn and——" he advanced a step.

"Lanty!" said Mabella, and in an instant he was grave.

"Dear girl," he said, "you don't think I would do anything to make you feel badly?"

The warning shriek of the whistle came to them.

"See, tie this round my neck, will you?"

She folded it with an adorable air of anxiety and precision, and stood on tiptoe to lay it on his shoulders and again on tiptoe to knot it under his chin, a process Lanty rendered arduous by putting down his chin and imprisoning her hands, a performance he found most satisfying. But at length he was off, and Mabella watched him round the corner of the barn, and then went indoors to attack the chaos upon the table with a good heart.

"Where's Vashti?" she asked.

"Spooning her young man in the garding," said Sally, emerging from her shell.

"Of all the impses I ever see!" ejaculated Temperance. "G'long and fetch in some wood." Sally departed.

"Vashti's in the garden peeling apples for supper," continued Temperance to Mabella, with an attempt at unconsciousness. Mabella gave her a hug.

"It's a sugar plum for Mr. Martin because you were bad to him, isn't it?"

"Yes, Lanty's had his——"

Mabella blushed, and an irrepressible ripple of laughter broke from her.

"Well, you needn't laugh," said Temperance. "Mr. Martin thinks Vashti's just about right. Well, there's no accountin' for taste. 'Everyone to their taste,' as the old woman said when she kissed her cow."

"Temperance!" said Mabella, "you don't mean——"

Temperance nodded oracularly, "Nathan thinks so too."

"Well!" said Mabella, and relapsed into silence. Here was news for Lanty. If Nathan and Temperance thought so it must be so. A fellow feeling not only makes us kind but often very acute; and in all Dole there were no such keen eyes for any "goin's on" (as courtship was disrespectfully designated) as those of Temperance and Nathan.

"Love, it is a funny thing;
It puzzles the young and the old;
It's much like a dish of boarding-house hash,
And many a man gets sold."

Sally's falsetto voiced this choice ditty with unction, as she entered with an enormous load of wood in her thin arms. She deposited the wood with a bang.

"Sakes!" said Temperance. "I wonder if she sings them songs to the preacher?"

Whereupon Sally, in vindication of her judgment, began a lugubrious hymn.

"Stop it," said Temperance. Sally stopped.

Beneath the trees Vashti peeled her apples busily, the narrow parings of the greenings twined about her white wrist, the thin slices fell with little splashes into the bowl of water which was to prevent them turning brown before being cooked. Miss Tribbey's apple-sauce was always like white foam. A voyaging wasp came, and settling upon the cores was very soon drunk, so that he was an easy prey to a half dozen ants which wandered by that way. The distant buzz of the threshing mill filled the air with a drowsy murmur as if thousands of bees hummed above a myriad flowers, here and there a thistle-down floated glistening in the sun. The scent of the overblown flowers mingled with the odour of the apples.

"Are we done now?" asked Sidney, as she laid down the knife.

"We are," she said with meaning emphasis. "Do you feel very tired after your exertions?"

"Not so tired as you'd imagine," said Sidney. "The truth is I couldn't bring myself to offer my services, for if you had accepted them I would have had to look at the apples instead of at you, and I did not have strength to make the sacrifice."

"Could you make sacrifices?" she asked.

"Try me," he half whispered. There was a tense moment. Mabella's voice came ringing from the house, the whirr of the threshing mill suddenly seemed near at hand, and through it there came Lanty's voice shouting some directions to the men on the stack.

"Perhaps I may some day," she said.

"You know," he said, his voice enchaining her attention even as she strove with bitter thought, "you know you will have the opportunity to ask anything, everything of me."

"Ah, how should I know?" she said, as one who has not deigned to observe too much.

Sally, sent out for the apples, appeared round the corner of the house.

"Promise me," said Sidney, "that you will come for a walk after supper; promise."

For an instant the boulders of Mullein meadow and the dimness of the twilight sky blotted out the crimson of the Virginia creeper on the porch which flamed in the sun.

"I will come," she said.

"Ah——" he said no more.

"Sorry t'interrupt," said Sally genially, as she stood beside them. "But painful as the duty is it must be did; but don't mind me, I'm blind in one eye and can't see out of the other."

"Sally," said Sidney very gently, "you talk too much."

For the first time in her life Sally blushed, and gathering up the apples and the parings departed abashed.

"You are not going in?" he said rising as Vashti stood up.

She held up her hands. "I must wash my hands," she said, "and I want to rest a little."

The slightest hint of fatigue or illness in the splendid creature before him always touched him strangely. It was like a sudden assertion of the human in something divine.

"Do," he said; "and Vashti," using her name with happy boldness, "you won't forget your promise."

"I never forget," she said, simply and sweetly.

He stood bareheaded watching while she entered. Then looking about, he suddenly noticed that in the garden the summer flowers were overspent, the little battalion of ants tugged viciously at their victim, whose yellow and black had shone so gallantly in the sunlight as he lighted down to sip the apple juice. The whirr of the threshing machine made melancholy cadences which sighed through the trees; and all at once the whole scene darkened.

It was only that the sun had dipped beyond the house, and the crimson Virginia creeper seemed in the shadow to be more brown than red, two or three of its leaves fell desolately to the earth, as dreams die when hope is withdrawn.

And Sidney, with the fatuity of lovers, said, "She has taken the glow with her."

But the torch which lighted Vashti Lansing's way was not filched from flowers and sunshine, but shone fed with the evil oils of anger and revenge, baulked will and disappointed love.

CHAPTER VII

The grey of twilight was paling the gold of the after-glow. A quiet hush had fallen upon the earth—rather intensified than disturbed by the lowing of far-away cattle. It was the quiet of raptured anticipation, as if great hands held the earth up to the baptismal font of the heavens to receive the chrism of night; and the earth, like a wise and reverent child, waited with hushed heart-beats for the benediction.

Sidney Martin waited in the porch for Vashti to keep her tryst, and presently he heard her footsteps. The echo of each step gathered in his heart, dilating it with happiness as an already full glass is brimmed above the brink by drop after drop. From his position, where he stood spellbound, he commanded an angled vista of the stairs, and slowly she descended within his range of vision; first the beautiful foot, proportioned so perfectly to the body it bore, then the long exquisite lines from heel to hip, and the yet more exquisite curve from hip to shoulder, and the melting graduation of breast to throat, and then the perfect face of her. She paused for a moment upon the last step, as if loath to step out of her pure rarefied atmosphere of maidenhood into the air vibrant with the sobs and sighs, the hopes and despairs, the gains and losses of human life; and standing thus, for one fleeting second there rose before Vashti a vision of renunciation. She saw herself, lonely but clad in righteousness, going on her way; but the next instant the austere dream vanished, brushed aside by a hateful, sneering cynicism. With a heart full of self-mockery, more evil than her evil intent, Vashti took the step to Sidney's side, and stood there the typification, as he thought, of gentle dignity and dignified womanhood.

"How good you are," he said gently.

They took the way almost in silence. She wondered vaguely where he would take her, to the far-away pastures, the little knolls nestling upon the hills which he loved, or to the oak trees where they had talked in the morning. When they reached the road she submitted her steps to his guidance with outward meekness and inward indifference. He turned away from Dole. It was to be the far-away pastures then—as well there as anywhere. But he had passed the gate! And then it dawned upon her. He was taking her to Mullein meadow!

Her indifference fell from her like a rent garment, bitter remembrance tore at her heart. How dare he bring her here and bid her masquerade amid these grey boulders where she had known such agony! She imagined those implacable rocks rejoicing in her humiliation. Were not her own curses yet hissing across the eerie barrenness of this wide waste field? Ah, even so, Vashti—if our curses do not seek us out we ourselves return to their realm; there is great affinity between a curse and the lips which utter it. The flame of her resentment fluttered to her cheeks, giving them an unwonted touch of rose. As they reached the entrance to Mullein meadow, she half stumbled; she recovered herself quickly, Sidney's swift touch being hardly needed to restore her poise.

To Sidney, her silence, the strange, sweet colour in her cheeks, her uncertain step, pointed but to one thing—the natural agitation of a girl about to have a man's love laid at her feet.

Surely never man was so exquisitely befooled as this one?

He took the path straight for the little spot where that happy betrothal had taken place. Vashti hesitated—this was too much.

"I—," she opened her lips to speak, but the words died away, unmerciful resolution freezing them at their source.

"Come," urged Sidney with tender insistence, and with an appearance of sweet submission she yielded, and at length they stood where those others had stood. The same grey sky bent above them, the same quiet hush brooded over the desolate reaches, the same clear star hung scintillant in the sky, and Sidney, taking her hands, which trembled by reason of the terrible restraint she was putting upon her anger, began to speak—very gently, but with an intensity which made his words instinct with life and love.

"You know," he said, "why I have asked you to come out to-night, but you cannot know why I have brought you here to this spot? It is because it is a place of happy auguries. Here, not knowing whither I strayed, I came upon the betrothal of Lanty and Mabella. Here, heartsick with envy of their happiness I turned away to face the desolate greyness of the twilight. Here I saw a star, one lone star in the grey, which seemed to promise hope, and in my heart I named it Vashti. See—there it is, but more golden now, more full of beneficent promise, burdened, as it seems to me, with gracious benediction. Oh, Vashti, when I left those two in the solitude of their happiness you cannot dream how my heart cried for you. All the way home nature's voices whispered in my ear 'Vashti—Vashti,' and my heart responded 'Vashti,' and it seemed to me that there was no other word in all the universe, for in it were bound all meanings. It seemed to me there was no other idea worth comprehending but the identity behind that word. Vashti, say that you love me—that you will marry me. Here, where my heart knew its bitterest longing, satisfy it with one syllable of your voice. Let me also build tabernacles here as the holy place where happiness descended upon me"; he let fall her hands. "Vashti, you know that I love you; give me your hands in symbol of yourself as a free gift."

He held out his hands. Slowly, gently, trustingly, as a woman who knows well what she does, and will abide by it, Vashti Lansing laid her hands in his. His vibrant, slender fingers closed upon them. There was an instant's pause——

"You love me!" he cried, as one, after a long novitiate, might hail the goddess unveiled at last. Then drawing her to him he kissed her on the mouth, and from that moment was hers—body—and yet more terrible bondage—mind; and she, with an astute and evil wisdom, forbore to make any conditions, any demands, till he had tasted the sweets of her acquiescence.

Would any man give her up, having held her in his arms, having touched her lips? With shameless candour she told herself, No. So she rested her head upon his shoulder, whilst he whispered in her ear the divine incoherences of love, and intoxicated with the charm of the woman in his arms, touched the white throat by the ear, where a curl of dark hair coiled like a soft, sweet shadow. A long, contented, yet questioning sigh came to him—

"Tell me?" he said.

"You will let me live always in Dole?" she said.

"Always—always, dear one! In Dole or anywhere else you like."

"Ah!" she said in a tone of dreamy happiness—"you will take old Mr. Didymus's place; we will live in the parsonage; what a happy life we will have!"

"Vashti!" said Sidney, almost reeling before the shock of her words. As a beautiful white mist rolls back to show some scene of sordid misery, so the glamour of the last few weeks lifted, and displayed vividly to Sidney all the awkwardness of the position which he had created for himself. Ever since that day, when stung by Sally's impertinent words he had agonized alone upon the hillside, nothing whatever had transpired to awaken its memory. A deference rather more pronounced than necessary upon the part of the village-folk, a certain constraint upon the part of the young men, had been the only visible signs that Dole remembered. But upon the other hand nothing had occurred which gave him the opportunity of explaining to Vashti, nor, indeed, had he ever been able to decide how he could explain to her, even if given the opening. He had gone to church with the Lansings Sunday after Sunday. Under the circumstances any other course would have been an insult to the *régime* of the house in which he was staying. He had found nothing in the little church which jarred upon his tastes or revolted his principles. The simple, pious sermons of grey-haired Mr. Didymus were entirely inoffensive to anyone not of *malice prepense* irritable. The sad experiences of his long life had mitigated his judgments. The man who in his fiery youth scoffed at death-bed repentances now spoke feelingly of the thief on the cross; the elect murmured among themselves that Mr. Didymus was "growing old and slack." Certainly his sermons were not learned, but neither were they devoid of a certain eloquence, for the old man knew his Bible by heart, and above all, they were free from the anecdotal inanity; it would never have occurred to the old, plain-spoken man to stand in his pulpit telling his people tales suitable for the comprehension of three-year-old children. There was, perhaps, the merest trace of asceticism in Sidney Martin's nature, and the simple doctrine of these people, their fatalistic creed, their bare little church, appealed to him as no gorgeous ritual or ornate sanctuary could have done. The hoarse, untuneful singing of these country folk, taking no shame of their poor performance, so that it was in praise of God, stirred his spiritual sympathies more profoundly than any cathedral organ—yet—he was a creature of reason, and he had always considered the Catholic Church more logical than any other, and above all, he had no belief whatever in the Christian doctrine. Ruled by a pure and lofty ideal of Truth, his life had been ideally good. His lofty aspirations did not lift him beyond sympathy with his fellows, only above their vileness. He adored nature with an almost heathenish idolatry, and had such reverence for her slightest manifestation, that he never willingly broke a leaf or crushed an insect. Literally, he worshipped the works, but not the Creator. And lo!—here was the woman round whom his very soul twined, taking it for granted that he believed all she did, and that his life could compass no higher happiness than to preach this belief to others; and what excellent grounds he had given her for thinking thus! All these things mirrored themselves in his mind in an instant, then he said:

"But, Vashti, I have no need to do anything. There are many worthier men than I to fill Mr. Didymus's place. I am not a preacher, you know."

"Oh, but you will be for my sake," she said, and laid her head down again upon his shoulder like a child who has found rest.

Truly there are more tempting devils than the urbane gentleman of the cloven hoofs.

"What had you meant to do?" she asked.

"Indeed, I had mapped out no definite course," he answered. "My mother's money makes life easy for me, you know, but I had meant to do something, certainly. Only I was taking my time looking about. I didn't want to do anything which would cut some fellow who needed it out of a living."

"Let me decide for you," she murmured; the breath of the words was warm on his ear. "Think how happy you could make us all. They all think so much of you in Dole on account of your prayer. Mary Shinar says you are a saint." Then, her arms stealing about his neck, she added, "Sidney, for my sake you said you would sacrifice anything. I didn't think this would be a sacrifice. I thought it would be a delight; but if it is a sacrifice make it for my sake."

Alas, he had fallen among the toils! He took swift illogical thought with himself. He would preach to them a pure and exalted morality. He would be the apostle of nature's pure creed. He would make Dole a proverb in all New England. He would teach, he would have a library, he would marry Vashti.

Glamoured by his love and his sophistry, his judgment, his sense of right and wrong, failed him. Sidney caught his Delilah to his heart.

"It shall be as you wish, my sweet," he said; "and now tell me you love me."

"I love you," she said, repressing the triumph in her voice. "I love you and I am proud of you," she said again, holding her head high. If she had lost much in Mullein meadow she had also gained a triumph there.

The short American twilight was darkening to night. The weird old boulders sentinelled round them might have been a druidical circle, and she the priestess fulfilling the rites. Nor was the victim wanting; only instead of slaying the body with a golden knife she had killed the soul with silvery speech.

"Ah," said Sidney as they turned to thread their way out of Mullein meadow, "surely this place is holy."

She paused, looking at him—"Do you not think that suffering sanctifies more than joy?" she asked.

"No, not such joy as ours, as Lanty's and Mabella's."

"I don't know," she said.

"But I'm sure of it!" he answered; then with a lover's fantastical fondness he went on, "I would not be surprised if when we visited this spot again we found it hedged in by lilies, tall white eucharist lilies, set to keep others from straying into consecrated ground."

"Sidney," she said, "promise that you will never, never ask me to come here again—it is too sacred."

He was deeply touched by her delicate, sensitive thought.

"Dear heart," he said, "never; yet do not the most reverent lips approach the sacramental cup more than once?"

"You will make a capital preacher," she said, "but you must not persuade people to do things against their conscience."

"You shall do as you like always."

They were on the highway by this time; a waggon overtook them, and then went on at a foot pace just in advance.

Vashti seemed to walk with intentional swiftness.

"Vashti," he whispered, "don't walk so fast. Let those people get out of sight."

"We must go on," she said.

Sidney thought this touch of shyness adorable in her who was so self-poised, yet he protested with zeal. Do men always try to destroy what they admire?

Suddenly Vashti bethought herself that an extra rivet was never amiss when one wanted bonds to hold, so with a sigh as of timorous yielding, she gave him her lips again in the shadow of the porch, and left him with a glory of happiness bedimming his mental vision.

The house was dim-lit and silent. After the labours of threshing-day everyone was worn out. Lights glimmered in the bedrooms, but the living rooms were dark.

Sidney paced up and down the little garden path for long, feeling "caught up to heaven, yet strangely knit to earth."

Vashti sought her room, and pulling up the blind looked out where Mullein meadow lay.

"A holy place!" she said to herself. "I wish I could pile the fire to burn all three of them. 'A tabernacle,' he said; I wish I might build me an altar there and slay them on it! I don't think even an angel would stay my hand. 'A sacrament'; I wish I had the filling of their cups, wormwood should they drink and the waters of Marah down to the very dregs—all three!"

Her nostrils dilated like a brute's upon traces of the prey. In the breast of such a woman love denied turns to gall. She paced up and down, up and down—her rage lent expression in grotesque gestures and evil words,

words which with Vashti Lansing's teaching and training she was superbly brave to use. It grew very late; her eyes were almost wild. She took the guttering candle in one hand and crept along the passage to Mabella's room. She opened the door and went in. Mabella lay asleep, her candid face budding from the prim little frill like a flower from its calyx. Vashti bent above her a haggard and violent face distorted by passion. Her eyes blazed; her lips drawn tensely back showed the strong white teeth. She leaned over the sleeper, her strong fingers closing and unclosing; a long tress of her hair fell across her shoulder suddenly and touched the dreamer's cheek—Mabella stirred, raised her hand half way to her cheek, murmured with a little happy smile—"Lanty—Lan—" her voice died away; her soft regular breathing continued unbroken. At the sound of that name uttered thus a dreadful purpose lighted Vashti's eyes. The fingers of her strong hand opened wide and advanced themselves toward the white throat which pulsed upon the pillow; at that moment the guttering candle fell over. Its burning wick and melted grease struck the hand which held it. Vashti instinctively uttered a smothered cry and jerked her hand; the light went out. Mabella stirred; Vashti sped to her room and got the door closed just as Temperance came to her door and said:

"Did anyone call?"

There was no response.

"Are you all right, Mabella?" she said, going across the hall to Mabella's door.

"Yes," said Mabella sleepily. "I think I knocked something over with my elbow and the noise woke me up."

"Are you all right, Vashti?"

"Yes, what is it?" answered Vashti.

"Nothin'—thought I heard a noise."

For hours Vashti Lansing lay and trembled with the only fear she knew; the fear of herself. How near she had been to terrible crime, only she and Omnipotence could know. She reflected upon consequences and told herself that never again would she give herself such an opportunity. At last she sank to rest, to be tormented till dawn by a strange vision.

It seemed to her she stood again in Mullein meadow, within the circle of boulders, and that slowly, slowly they closed in upon her; closer and closer they came, narrowing about her with gradual but horrible certainty, and at last they touched her and held her tight, shackling her hand and foot so that she could not move a muscle, but they did not kill her; and whilst she was thus held all Dole defiled before her; the villagers pointed at her with scornful fingers and passed whispering on; her mother, who had long been dead, passed with her father, but they did not look at her, nor seem to know she was there, nor did old Mr. and Mrs. Didymus who presently joined her father and mother. Then the scene grew brighter and she saw Temperance and Nathan together; they shook their heads, looking at her sadly but coldly; then a sweeter radiance flooded the view upon which she looked, and Mabella and Lanty with their little children about them drew nigh her, and they spoke kindly words to her, and put a shade over her head to keep off the sun's heat, and raised a cup to her lips, and one of their children came and held up a child's haphazard bouquet to her nostrils that she might smell the flowers. She tried to repulse these kindnesses; she tried to drive Mabella and Lanty away with evil words, but the stones pressed too tightly upon her to admit of speech, and while she writhed thus impotently, she looked far away where one wandered alone; there were butterflies and birds about him, and flowers springing about his feet, and he wore a look of calm ineffable happiness, and, yet, it was not the same happiness as shone upon the faces of Lanty and Mabella which lighted the eyes of this visioned Sidney. But in her dream Vashti did not dwell long upon this, her thoughts reverting to the paralyzing prison which encompassed her; and she fought, and struggled, and strove, yet she could not move those terrible stones, and casting her eyes down upon herself, it gradually dawned upon her that she could not even struggle. The terrible wrenches and efforts she had made were but imaginary, so tightly was she held that she could not so much as twitch a finger. Thus the hours passed with her.

Mabella slept sweetly and healthfully, so rapt in love that even the baleful influence bending over her so terribly in the night had had no power to disturb her rest, although the gaze of even a friendly pair of eyes so often murders sleep. Sidney slept also, and high above the pale wastes of Mullein meadow, the star of promise still shone, unrecking of the presumptuous human heart which had dared to dream its silvery splendour a pennon of hope.

CHAPTER VIII

When Sidney opened his eyes next day it was upon a transfigured world that he looked. A world golden with imaginings of happiness across whose vistas shone a white path, like the milky way in the heavens, marking the life road to be trodden by Vashti and himself. Cradled in a happy trance his heart knew no apprehensions. At such a time retrospect shares the mind almost equally with anticipation. The glorious present is made still more glorious by comparison. As Sidney dwelt upon his past it was borne in upon him with peculiar force that it had been but a curtain raiser to the real drama of his life. He had been a popular man as a student and afterwards also, but it seemed strange even to himself how few real ties he would have to sever in adopting this new life—so radically different in vocation from any he ever dreamed of before. The fact was that in all his friendships he had given more than he had received. He had given liberally of that intangible vital capital called sympathy, and he had received but little in return. Although he had not realized it his friendships had been only so many drains upon his vitality. He had thought of, and for, his friends continually; they had accorded him the tribute of uncomprehending admiration which bears the same relation to real sympathy as bran does to the full, rich wheat. Thus it was that in separating himself from these friendships he felt no wrench. Separate from them he must. He knew that the keeping of his promise to Vashti was utterly incompatible with his old life; he must "come out and be separate" from all his old associates and associations. He felt, however, that this would be possible; possible without sacrilege. His attitude towards religion had always been defensive rather than offensive. He felt deeply the pathos of the Christ drama. The figure of the Man of Sorrows was a familiar one in the gallery of mental portraits to which this idealist had turned in time of trial for strength.

There was one man whose verdict upon his action he longed to know, yet dreaded to ask. A strong soul, untamed by sect, unshackled by formulated belief. A man whose magnificent active human organism was hallowed by the silver thread of mysticism. A man whose splendid logical mind was transcended by a subtle sense of premonition, intuition, which led him far beyond where his reason or his scanty learning could bear him company. A man whose eyes looked out wistfully yet eagerly from beneath penthouse eyebrows. A man whose toil-roughened fingers turned reverently the pages of books he could not read; French or German books beyond his ken. A man in whose proper person Sidney had always felt there was symbolled forth the half blind, half perceptive struggle of the human to comprehend the infinite.

What would this man think of his new vows? This man who would have died for what the world called his *disbeliefs*.

Well, Sidney told himself that his first *devoir* was to Vashti and the promise made to her. He would not delay. These thoughts bore him company till he was in the hall. He did not know the hour, but suddenly he was aware of a subtle, penetrating freshness in the air. He looked out of the hall door: the garden was dim with autumnal dew. Was it indeed so very early?

He heard voices in the kitchen. He found there only Mr. Lansing and Miss Tribbey.

"Is it so early?" he asked, smiling.

"For the land's sakes! Mr. Martin!" said Temperance. "Is that you?"

Sidney laughed aloud; there was a ring in his voice which made Temperance regard him.

"I have been awake for ages," he said; "so here I am."

Temperance remembered certain days in the past when she had been wont to awaken ere the first bird sang in the dark. These were the days when Nathan, a hobbledehoy, too bashful to woo her in daylight, used to waylay her in the lane when she took the cows back to the field, and stand with his arm about her in the dusk.

Temperance rubbed her eyes.

"The morning sun do dazzle," she said, giving unsought explanation of the moisture in her eyes.

"Better set right down and have breakfast," said old Mr. Lansing. "The young folks is turrible lazy, it seems to me, nowadays."

"Oh, not all of them," said Sidney. "Look at Temperance."

Old Lansing chuckled delightedly.

"Nathan Peck had better look out, Temp'rins; I allus did say you had a way with the men."

Temperance tossed her head, well pleased.

"Will you have your eggs fried or biled?" she asked Sidney, the blush upon her gaunt cheek giving her a sadly sweet look of girlhood.

Old Lansing finished his breakfast and pushed back his chair.

"You'll excuse me," he said, "but I've been up since cock-crow, and I hav'n't done a blessed thing but water the cows. The men are in the barn now waiting. Temp'rins 'll give you breakfast. I'll warrant the girls will be surprised when they get down. Lazy critturs! Temp'rins, why don't you wake 'em up?"

"O sakes! Let 'em sleep," said Temperance; "in a few more years they'll wake fast enough o' their own accord. Laws! I kin mind when I'd have slep' all day if they'd let me be."

In this homely sentence lay the secret of Temperance's influence. This gaunt old maid never forgot the workings of her own youth. Indeed now that it was past she acknowledged its weaknesses very frankly, and this reminiscence made her very lenient towards young people.

Old Mr. Lansing departed for the barn, and Sidney, filled with impatience to see Vashti, paced up and down the kitchen.

Temperance brought the eggs and sat down beside the tray, looking at him with a sort of pitiful sympathy in her keen eyes.

Sidney essayed to begin his breakfast; a smile twitching the corners of his sensitive mouth.

Temperance watched him.

At length he laid down his knife and looked at her.

A subtle atmosphere of sympathy made him confident and expansive.

"I say, Temperance," he said, "I was never so happy in all my life. You don't mind my talking to you about it, do you? I'm so happy that—oh, Temperance."

It was a boyish conclusion; he looked at the gaunt country woman; her hands worked nervously; she looked as if she *felt* the emotion which made him ineloquent.

"You have seen—you are pleased?" he continued in haphazard fashion.

"Bless your soul, Sidney," burst out Miss Tribbey, forgetting to be formal, "I'm pleased if so be you're happy. I ain't very religious. I expect I have a worldly heart. I'm like Martha in the Bible, allus looking after cooking and sich, but I've said to my Nathan heaps o' times, 'He's a blessing,' I said, 'to have in the house,' and I mean it. My soul! I only hope Vashti 'll come up to your expectation."

"Ah," said Sidney, "there's no doubt of that. She's perfect."

Miss Tribbey's mouth half opened, then shut resolutely. She had her own standard of perfection, but she had too much sense to deprecate the lover's fond extravagance.

"I'm perfectly content," said Sidney, "perfectly."

Miss Tribbey grew very white.

"Don't say that," she said earnestly, "don't; no good ever came of sich a boast. It's terrible dangerous t' say you are perfectly content. I never knew good to come of it—never."

"But I am," said Sidney, feeling happy enough to challenge the powers of evil *en masse*.

"Listen," said Temperance gravely, "don't say that. 'Taint meant for mortal man to be content. 'Taint intended. What would make us work for Heaven if we was perfectly content here? No, don't say it. I've known one or two people that thought themselves perfectly content, and how soon they was brought down! There was Mrs. Winder. Has anyone told you about Mrs. Winder?"

"No," said Sidney, "but I know her by sight. She's got a stern face."

"Starn! You'd be starn-looking too if you'd come through what Sal Winder has. First she married Joshua Winder; he was a bad lot if ever there was one, and after they'd been married ten years and had four children, what does he do but up and run away with a bound girl at Mr. Phillipses, a red-cheeked, bold-faced critter she was. Well, Sal never said nothin'. She was left with a mortgage and the four children and a roof that leaked. I don't s'pose anyone ever knowed the shifts Sal was put to to bring up them young ones and work that place and make both ends meet and keep the roof of the old house from falling in. Mebbe you've remarked the old house? It's got a white rosebush by the door, and blue ragged-sailors in the yard, and the pile of bricks beyond was once a smoke house. She had all her hams and bacon stole one year to make things easier for her. Well, her oldest boy was the most remarkable young one that Dole ever see. Joshua his name was, after his father, but that's all the likeness there was between the two of them. That boy was jist grit and goodness clean through! And the way he helped his mother! There wasn't a foot of that old place they didn't work, and prices were good then, and in about six years Sal got the mortgage paid. She gave a dollar to the plate in church the next Sunday. Some held 'twas done to show off, but Sal wasn't that stripe of woman. 'Twas a thank-offering, that's what it was.

"Well, next year Sal built a barn, and the year after the new house was begun. The house went on slowly, for Sal wanted to pay as she went along.

Well, at last the house was built and painted real tasty, and one day I was over there to visit a spell, and Sal says, 'Joshua has gone to pay the painter for the house painting,' she says; 'it's a sort of celebration for us and we're having ducks for supper. I hope you'll stay and help us celebrate.' Then she went on to say how good Joshua had been, which she didn't need to tell me, for all Dole knowed he was perfect if ever there was a perfect son. So jest after the lamps was lighted, in come Joshua. He was tall and slim; he favoured Sal in his looks; he had worked so hard ever since he was little that his hands had a turrible knotty look like an old man's, and he had a sort of responsible expression to his face. Well, we was all setting at supper and Joshua had cut up the ducks and we was all helped, and Sal says, 'Now make your supper all of ye. We've had a hard row to hoe, Joshua and me, but we've kep' it clear o' weeds, and I guess we're goin' to have a harvest o' peace and quiet after the grubbin'.' Joshua looked up at his mother, and I never seen two people more happy to look at. Sal was real talkative that night, and she says:

"'Well, Temperance, I'm right glad you're here to-night. *I'm perfectly content this night*,' she says. The words wasn't out of her mouth till I saw Joshua give a shiver—like a person with a chill in his back.

"'Have you got a chill, Joshua?' I says, and he laughed quite unconsarned, and he says, 'Yes, I seem to have the shivers.'

"Four days after that Joshua Winder lay dead in the new house.... My! I mind how his hands looked in his coffin. His face was young, but his hands looked as if he'd done his heft o' work. No, never say you are perfectly content. It's turrible dangerous."

Sidney's sensitive heart was wrung by the homely story.

"Oh, Temperance," he said, "why did you tell me that?" She looked at him as a surgeon might regard one whom his healing lancet had pained.

"Because," said Temperance, "because it's a tempting o' Providence to say or to think you are content. I ain't superstitious, but I'd rather hear the bitterest complainings as to hear anyone say that."

"And yet," said Sidney, "I should think the Lord would be pleased to see people happy, each in his own way."

"Well," said Temperance, modestly, "I ain't much on religion, Mr. Martin. I can't argue and praise and testify the way some can, but my experience has been that when folks begin to think themselves and their lives perfect and to mix up earth with heaven, and forget which one they're livin' in, they're apt to be brought up sudden. It seems to me heaven's a good deal like a bit o' sugar held in front of a tired horse to make him pull.

I guess there's a good many of us would lie down in the harness if it wasn't for that same bit of sugar; we may look past the sugar for a while, but when we get to a bit of stiff clay or run up against a rock we're mighty glad to have the sugar in front o' us again; but, sakes! you ain't made no breakfast, and there's the girls! You'll breakfast with—her—after all."

Temperance gave him an arch look and departed, and Mabella had hardly crossed the threshold before the sympathetic Miss Tribbey called her; when she arrived in the back kitchen Temperance took her by the shoulders and whispered energetically in her ear:

"Sakes, M'bella! Don't go where you ain't wanted."

Mabella's eyes lighted with sympathy.

"You don't say!" she said.

Temperance nodded like a mandarin.

"It must be catching!" said Mabella. "It was Nathan brought the infection to the house."

"Go 'long with you," said Temperance, and with a very considerate clatter of dishes she made her intended entry audible to the two people in the kitchen.

Mabella looked at Vashti eagerly—sympathetically, but the calm, beautiful face of her cousin was as a sealed book.

"Whatever was that noise in the night, Temperance?" asked Vashti.

"Why, I don't know," said Temperance, "I was sure I heard a noise, but I couldn't see anything when I got up. Did you hear anything, Mr. Martin?"

"Not I," said Sidney, "but I was so busy with my own thoughts that you might have fired a cannon at my ear and I would not have heard it." He looked at Vashti; her down-drooped eyes were fixed upon her plate; suddenly he exclaimed:

"What have you done to your hand? It's burned!"

"Yes," she said quietly, "after I blew out my lamp last night I knocked the chimney off. I caught it against my side with the back of my hand, that burned it."

"My!" said Mabella. "I would have let it break."

Vashti smiled, and suddenly raised her eyes to Sidney.

"A little pain is good for me, I think. It makes one know things are real."

"But the reality is sometimes sweeter than the dream," he said, tenderly.

She let her eyes fall in maidenly manner. It was as if she had spoken. This woman's most ordinary movements proclaimed the eloquence of gesture.

"You must have been up early," said Mabella to Sidney.

"Yes," he said, "I was in a hurry to leave the dream-world for the real."

"And how do you like it?" asked Mabella, saucily.

Vashti spoke at the moment, some trivial speech, but in her tone there was the echo of might and right. It was as if with a wave of her hand she brushed aside from his consideration everything, every person, but herself.

They rose from the table together.

"Come out," he whispered; she nodded, and soon they were pacing together in the morning sunshine. Mabella looked after them; turning, she saw Temperance wiping her eyes.

"What is it?" she asked with concern.

"Nothing," said Temperance; "nothing; I'm real low in my spirits this morning, though why, I'm sure I can't say. But it's fair touching to hear him! There he was this morning talkin' of her being perfect, and sayin' he was perfectly contented. It's a tempting o' Providence. And, Mabella, there's Vashti—she—well, I may misjudge," concluded Temperance lamely. "Sakes! look at them chickings," with which Temperance took herself off to regulate the ways and manners of her poultry yard. Mabella departed to do her work light heartedly, and Vashti out in the morning sunshine with her lover was weaving her web more and more closely about him.

In two nights more Sidney was leaving Dole.

It was the night of the prayer meeting.

All Dole knew of his engagement to Vashti Lansing; all knew he hoped to be the successor of old Mr. Didymus. The old white-headed man had spoken a few words to him telling him how happy he was to think of his place being so filled. He spoke of it calmly, but Sidney's lips quivered with emotion. Mr. Didymus said, "Wait till you're my age and you won't think it sad to talk of crossing over. Wife and I have been two lonely old people for long now, hearkening for the Lord's voice in the morning and in the evening, and sometimes inclined to say: 'How long, oh, Lord! How long?' We won't be long separated. When folks live as long together as we have they soon follow each other. That's another of God's kindnesses."

There was in the simple old man's speech an actual faith and trust which brought his belief within the vivid circle of reality.

"I will do my best," said Sidney.

"The Lord will help you," said the old man.

The prayer meeting was animated by thought for Sidney. There was something in the idea of his going forth to prepare to be their pastor which caught the Dole heart and stirred its supine imagination.

When old Mr. Didymus prayed for him, that he might be kept, and strengthened and guided, it was with all the fervour of his simple piety. The intensity of his feelings communicated itself to his hearers. *Amens* were breathed deeply and solemnly forth.

Vashti would have liked Sidney to speak.

"I cannot," he said simply; nor was his silence ill thought of. He was going forth; he was to be comforted; he was the one to listen to-night whilst they encouraged him and plead for him, and again, in the name of the Great Sacrifice, offered up petitions for him. The hour had come for the closing of the meeting, when suddenly Mary Shinar's clear, high treble uttered the first words of one of the most poignantly sweet hymns ever written.

"God be with you till we meet again—
May His tender care surround you,
And His Loving arm uphold you,
God be with you till we meet again."

Every voice in the church joined in this farewell, and then the benediction was slowly said—the old tender, loving, apostolic benediction, and they all streamed forth into the chill purity of the autumn night. They shook hands with him, and he stood among them tall and slight and pale, inexpressibly touched by their kindliness, unexpectedly thrilled by their display of emotion. It was only their religion which moved these people to demonstration.

The last hand clasp was given. The lights in the church were out, and the Lansing party took its way homeward.

Temperance's face and Mabella's were both tear-stained. Vashti's pale beauty shone out of the dusk with lofty quietude in every line.

Sidney looking at her felt he realized what perfection of body and spirit meant.

A new moon was rising in the clear pale sky—the wide fields, tufted here and there with dim blossomed wild asters, lay sweet and calm, awaiting the approach of night as a cradled child awaits its mother's kiss. Far away the tinkling lights of solitary farm-houses shone, only serving to emphasize the sense of solitude, here and there a tree made a blacker shadow against the more intangible shades of night. There was no sound of twilight birds; no murmur of insect life.

Sidney was passing home through the heart of the silence after a farewell visit to Lanty, who was kept at home nursing a sick horse.

It was the night before Sidney's departure from the Lansing house. The summer was over and gone. It had heaped the granaries of his heart high with the golden grains of happiness. He walked swiftly on, then suddenly conscious that he was walking upon another surface than the grass, he paused and looked about him. Around him was the tender greenness of the newly springing grain—above him the hunters' moon curved its silver crescent, very young yet and shapen like a hunter's horn. A new sweet night was enfolding the earth, gathering the cares of the day beneath its wings, and bringing with it as deep a sense of hopeful peace as fell upon the earth after the transcendent glory of the first day, and here amid these familiar symbols of nature's tireless beginnings he was conscious of an exalted sense of re-birth. He too was upon the verge of a new era.

He stood silent, gazing out into the infinity of the twilight.

Afterwards when the pastoral mantle did fall upon his shoulders there was a solemn laying on of hands, a solemn reception into the ranks of those who fight for good; but the real consecration of Sidney's life took place in that lonely silent field, where the furrows had not yet merged their identity one with the other, where the red clods were not yet hidden by the blades. Out of the twilight a mighty finger touched him, and ever after he bore upon his forehead almost as a visible light the spiritual illumination which came to him then. It was, alas, no self-comforting recognition of a personal God. It was only the sense that all was in accord between the Purposer and the world he had made; but this was much to Sidney. The man-made discord could be remedied, even as the harsh keys may be attuned. For ever after this hour he would give himself up to striving to bring his fellows into accord with the beautiful world about them.

Suddenly he felt himself alone. A speck in the vastness of the night, a little flame flickering unseen; but just as a sense of isolation began to fall upon him a mellow glow gladdened his eyes—the light from the open door of the old Lansing house. He bent his steps towards it with a humble feeling that he had trodden upon holy ground ere he was fitly purified.

In after days when many perplexities pressed upon him, he often withdrew in spirit to this twilight scene. Of its grey shades, its dim distances, its silence, its serenity, its ineffable purity he built for himself a sanctuary.

Alas! In that sanctuary the God was always veiled.

CHAPTER IX

It was nearly two years after Sidney went forth to prepare for the pastorate of Dole, when he stood one morning reading and re-reading the brief words of a telegram:

Come at once. Mr. Didymus is dying.

VASHTI LANSING.

The old man had been failing fast since the springtime.

The first April showers were quickening the earth when one day Sally found Mrs. Didymus dead in her chair, her Bible upon her knee, her spectacles pushed up on her brow, her dead face turned to where upon the wall hung a faded and discoloured portrait of Martha.

"It won't be long now," Mr. Didymus had said to Sidney upon that occasion, and Sidney felt it would be cruel to contradict his hope.

All summer long as Sidney read Vashti's accounts of the old man's fluctuating health he had thought of the solemn gladness of the moment when the summons should come. His loins had been girded for months past and now he was to set forth.

He had said to Vashti in a wistful letter, "When the hour comes be sure you send for me yourself. Let it be your personal summons which brings me to your side." And now such a summons lay before him.

He had no preparations to make. All that required to be done could be arranged afterwards. But, ere he set out for the new life, he had one visit to pay. He had always promised himself that when the hour came he would not taste of its joy till he had gone to the man of whom he had thought during the first gladness of his engagement.

Surely it was a curious thing that a minister of the Gospel should seek counsel of an unlearned agnostic. Nevertheless Sidney went confidently. At each step he took towards his destination he grew more and more ashamed for that he had so long withdrawn himself from this man.

Sidney found him in his old place amid the whirring wheels of the great factory in which he worked. His grizzled hair was a trifle greyer, his strong figure a little more bent; but his clear cut mouth was as firm as ever, his eyes as wistful and eager. They had that expression of receptiveness which so often marks the true sage, who, very wise, is yet always eager to learn.

Between the sliding belts Sidney encountered his delighted gaze fixed straight upon him. The visitor threading his way with difficulty through the maze of machinery to where he stood with such a welcome in his eyes that Sidney's impulse had been to brave the wheels and go straight.

"How I wanted to come and meet you," said the man, holding out a begrimed hand eagerly. "But you know my hand must be on the lever always."

"Ah," said Sidney, "I felt your welcome even before I saw you, and when I saw you formalities were discounted."

The man looked at him, a shade of awe solemnizing the gladness of his face.

"There are some things which almost frighten one," he said. "Do you know that all day long I have been thinking of you, remembering the lectures you used to give us at the Shelley Club and wondering if I should ever, ever hear from you again?"

"And now I am here!" said Sidney.

"Yes," said the man, looking at him lovingly. "And it is so good to see you."

In the midst of his happiness Sidney remembered to say "And how does the Shelley Club progress? Are you president yet?" The man shifted his feet awkwardly.

"Yes, I am," he said.

"Ah, the right man in the right place," said Sidney cordially. "So the club goes on?"

"Yes, we have nineteen members now and there are often fifty at the meetings."

"There's a stride!" said Sidney. "We used to be proud of ourselves if we could say 'we are seven,' didn't we? Well, I would like to hear your addresses."

"You have some news to give me, I am sure," said the man, who, during the conversation manipulated his lever with the mechanical precision of a man whom practice has made almost automatic.

Sidney flushed.

"Could you come out for a few minutes' quiet talk?" he asked.

"I shall see," said the man, turning a knob which arrested the wheels. He went to a man almost as grimy as himself, but who wore a coat. Sidney looked about him with shuddering disgust at the surroundings.

The machinery beside him shivered with the suppressed energy kept in check by the knob the man had turned. It seemed to Sidney a symbol of the eager soul of the man whom he had come to see, prisoned by circumstances within the circumference of petty cares, yet quivering and throbbing with divine energy.

The man was returning pleased with the little boon of time he had gained. The circumstances gripped Sidney's heart. He felt his own freedom and ease a reproach.

The man led the way, turning down the sleeves of his grey flannel shirt. He passed broad-shouldered between the whizzing belts, one touch of which meant mutilation. Sidney edged his way gingerly after him. The spaces between the whirling wheels seemed very narrow.

The workman led the way out into a desolate but sunny little courtyard. A high wall enclosed it; great heaps of packing cases filled one corner; a freight car, run in upon a little row of rails, stood just within the gate.

"Sit down," said the man, waving Sidney to a place upon a pile of boards. It struck Sidney that there was a sense of luxury in the way in which he let his frame relax; it was an unaccustomed treat, evidently, these few moments stolen in the midst of the sunshiny forenoon.

"Now for your news," said the man. "Is it about yourself?"

"Yes," said Sidney, "and it will surprise you greatly. I am about to become, in fact already am, a Minister."

"Of what—to whom—where?" asked the man.

"A preacher of the Christian gospel," said Sidney. "To a pious little community in the New England hills."

There was silence for a moment. The whirr of the wheels came to them, they heard a postman's whistle in the street outside and the chirping of some sparrows which fluttered about the empty car.

"You are disappointed," said Sidney; "you disapprove, but——"

The man raised his hand.

"It's for a woman, I suppose," he said. "Would nothing satisfy her but your soul?"

"Oh," cried Sidney, "I will do my duty by them. I will preach the truth to them. They shall know how noble and lovely life may be. They shall be shown what real beauty is, and told that righteousness for righteousness' sake is the highest good."

His friend sat silent still; Sidney looked at him almost pleadingly, and saw that his eyes were blurred by tears.

"Listen," he said to Sidney. "Give it up. You don't know what you are doing. It will kill you. I know you so well. You are salving your conscience now by good resolutions. When you see the fruitlessness of it all you will torture yourself with thoughts of your responsibility and what not, and the end will be chaos."

"Do you think I have not nearly gone mad already?" said Sidney, growing very white. "Surely you must guess how I have questioned my ability to do them good. But I think the worst of that is past now. I shall have a stay, a support, an inspiration which will never flag. The most beautiful and best woman in the world has promised to marry me the day I become minister of Dole."

"I've heard of the devil baiting his line with a woman," said the workman contemptuously, but yet in such a manner that Sidney could not take offence. Then he went on:—

"You say you'll do your duty by these people, but it's not that I'm thinking about. It's you. Remember this, you are to work in the vineyard of human nature, its soil is the shifting quicksand of human weakness. When you feel that sucking you down, to what will you turn? Upon what secret source of strength can you draw? Do you think the men who preach the Christ word in the slums could live and eat and continue their work unless they drew strength from some unseen reservoir? No, a thousand times no. Of course, I think their belief a delusion, but it is real to them, as real as the Divinity of Truth, and Truth alone, is to me. To preach a personal God without belief in one is to court destruction; at any moment, by disappointment or self-reproach, you may be thrown back upon your own beliefs. Shall the mother whom you have denied open her arms to you? Or shall the personal God in whom you do not believe sustain you? No, you will fall into the void. Sidney, give it up."

There was a pause.

"I will never give it up," he said. "I have promised that I shall devote myself to the work, and I will. You speak as if I had denied Nature and spat upon Truth. I have done neither. These two things will bear me through. There was one night in the fields—there was a new moon, and the young grain was springing. I saw things very clearly just then. I felt I could do good, and that it was my bounden duty to try. Bid me good-speed."

The workman rose. He took Sidney's hand and pressed it in both of his.

"I think," he said, "no human being ever began a hopeless course with more sincere and honest good wishes." As he held Sidney's hands and looked into the grey eyes of the younger man his own keen eyes dimmed and grew seerlike. The look of the visionary illumined his face.

"You will toil and strive and suffer," he said. "You will spend and be spent for others. You will have griefs, but you will never realise them, for you will be too absorbed in the sorrows of others to feel your own. You have bound yourself to a wheel, and until you are broken upon it, and your spirit spilt into the bosom of the Eternal, you will never know you have been tortured."

A half sob arrested his speech.

"Good-bye," he said, "good-bye!"

"Good-bye," said Sidney, who was much moved. So the two men parted. The one went into the sunshine; the other back into the hot atmosphere, where the deleterious dust was eddied into maelstroms by the whirling wheels.

The one murmured, "Vashti, Vashti"; the other, as he oiled the wheels and bent strenuously over his work, thought long and sorrowfully of many things. It chanced to be the meeting-night of the Free Thinking Vegetarian Club, of which he was president, and in his little speech he said much of a man who bartered his soul for a mess of pottage. But he told the story in such fashion that this man seemed to shine as an unselfish hero before their eyes, instead of as a weakling, spendthrift of a precious heritage of independence.

Thus an author has sometimes such wholesome charity for his villains that we love them more than their betters.

As Sidney was borne towards Dole that day, he relived as in a vision all the events which followed that first haphazard visit of his. And yet, could such a vital event be born of chance?

How well he recalled the peculiar fancy he had had when Dr. Clement, after his visit to the country, gave him old Mr. Lansing's invitation.

It was as if a little bell set swinging in his father's boyhood had suddenly tinkled in his ear, bidding him turn in his youth to those scenes where his father had been a boy.

He remembered the day when dear old Temperance first opened the door to him. He knew now the enormity of his going direct to the front door. In Dole only ministers and funerals went there. Sidney never really acquired the etiquette of the Dole doors. One has to be born in a court to properly appreciate its etiquette.

With epicurean delay the gentle stream of his recollections took him down the road, past Mullein meadow (O! place of promise!), to the "unction sale," at Abiron Ranger's, and then his memory leaped the bounds, swept aside intervening incidents, and dwelt upon the glorified vision of beautiful Vashti. Ah! "Who ever loved that loved not at first sight?"

Then followed his long visit with its rhythmic lapse of happy days.

Then, the Holy Grail of her heart had been won.

And afterwards came the long waiting. The short visits to Dole. And now!

The marriage of Mabella and Lanty had taken place a month or so after Sidney left Dole the first time. Their little daughter Dorothy was more than a year old now.

Temperance and Nathan were not yet married, but three months before Temperance had bought a new black cashmere dress in Brixton, and Nathan was known to have priced a china tea set, with gilt rosebuds in the bottoms of the cups. Dole felt, therefore, that matters were approaching a crisis with Temperance and Nathan.

Old Mr. Lansing had grown very frail. He had had a stroke of paralysis, and had never been the same man again. His eyes always had an apprehensive look which was very painful to witness; and strangely enough this quiet, self-contained old man, who all his life had seemed so content with the little village where he was born, so scornfully unconscious of the world which fretted and throbbed beyond its quiet boundaries, now showed a great eagerness for word from the outside. He subscribed to several newspapers. And when Sidney came the old man would question him with persistent and pathetic eagerness about the details of different events which he had seen chronicled with big typed headings, and Sidney found himself often sore at heart because he knew nothing whatever about the matter. American journalism has some grave flaws in its excellence, and surely the hysterical lack of all sense of proportion and perspective in presenting the picture of the times is a deplorable thing. It does grave and positive harm in the rural districts where it is impossible for the people to gauge the statements by comparison with events.

Sidney was greatly touched by the misconception of old Mr. Lansing in regard to these things.

"Ah," said the old man once, laying down the paper in which he had read a grotesquely exaggerated account of some political caucus which was made to appear like a meeting of the national powers, "Ah, there's no wonder dear old Sid went to Bosting." He shook his head and sat with his elbows upon his chair, looking before him into vacancy. What fanciful vista of possibilities did he look upon? What vague regrets beset his mind? To Sidney this was unspeakably pitiful. This old man with his young dreams—and it was the more sad, inasmuch as the dreamer himself knew their futility.

Old Lansing had always been a "forehanded" man with his work. He had never left over one season's duties till another, but he had forgotten to dream in his youth, and now he was striving in his age to overtake the neglected harvests of his garden of fancy.

When the train stopped at Brixton the first person whom Sidney saw was Lanty. Lanty tall and strong, and debonair as ever. He greeted Sidney very heartily.

"You've come for keeps this time," he said, as he led the way to where the roan, a trifle more sedate than formerly, stood waiting between the shafts of a very spick-and-span buggy.

"We will go straight to the preacher's," said Lanty. "I hope we'll be in time."

"Is he so low?"

"Dying," said Lanty simply. He touched the roan with the reins and it sprang forward. Sidney's heart fled before. The landscape upon either side stretched dimly before his eyes. He was conscious that Lanty was speaking to him, and he made suitable replies. But all his mind was glamoured by one thought, for Vashti had promised that Mr. Didymus should marry them.

Was this then THE DAY?

They passed Lanty's house, a square building with heavily timbered porch, and Lanty drew rein to call "Mabella, Mabella!" But there was no reply.

"She must have gone into Dole," said he, and once more they went on. Ere long they were driving up the streets of Dole. The women stood at the doorways with elaborate pretence of being occupied. The men endeavoured to infuse surprise into their recognition of Sidney, although most of them had purposely elected to stay in the village "choring" around the house instead of going to the fields or the woods.

The wise wives of Dole, knowing the amiable weakness of their husbands, had preferred special requests that day to have work done about the house. In Dole a man always thought he was conferring a personal favour upon his wife if he straightened up a leaning garden fence, mended a doorstep, or banked up a cellar for winter. There were six cellars banked up in Dole on the day when Sidney entered it. Upon the spring air the odour of fresh-turned earth speaks of new ploughed fields and fresh harvests, but in autumn the earthy smell is chill and drear, and brings with it a sense of mortality, a hint of the end. And this atmosphere hung heavy over the little village as Sidney entered it.

As the buggy drawn by the roan horse passed, the ranks of Dole closed up. That is, each woman crossed to her neighbour, and the men rested from their labours to discuss the arrival.

There was one thing that never was forgotten about Sidney's entry—a circumstance viewed severely by the many, leniently by the few—he wore a grey suit of clothes. Dole murmured in its heart at this infringement of the ministerial proprieties, but Dole was destined to experience a succession of such shocks, for its young and eager pastor trod often upon the outspread skirts of its prejudices.

Sidney himself was profoundly moved as he drove up the street, for he was entering the precincts of his holy city. In the geography of the heart there are many cities. There is the place where we were born; the place of our dreams; the Rome which under one guise or another fills the foreground of our ambitions; and above all there is the place where first we tasted of love, ah, that is where the Temple Beautiful stands. And Sidney's first and only love had been born in Dole.

Eager eyes were watching for them from the parsonage windows; Mabella, the habitual happiness of her face masked and subdued by tender-hearted concern; Mrs. Ranger, a bustling important woman of many airs and graces, filled with a sense of her own importance, and knowing that her every action would be reported to Temperance Tribbey (her sworn enemy) by Mabella; Mr. Simpson who had nursed Mr. Didymus from the beginning; and, waiting alone and silently in the tiny hall upstairs, Vashti Lansing.

She saw the two men coming up the street, side by side in the buggy, and her heart leaped up and cried for the one who was denied her. Again an angry gust of passion shook her as she looked. For the one moment her decision wavered. That pale slight man whose grey eyes were so eager, so alight with hope and love, was nothing to her compared to the blue-eyed, fair-haired young countryman. Why should she condemn herself to the torture of the continual contrast? But this way her revenge lay, unplanned yet, but so eagerly desired. She would surely, surely find some means to make them feel her power when as the preacher's wife she was First Lady in Dole. So Vashti Lansing, filled with Samson-like courage to wreck her enemies at any price, slowly descended the stairs as Sidney entered the front door. Then she went towards him.

Mabella saw them and with adroit sympathy endeavoured to detain Mrs. Ranger in the kitchen. But that worthy woman saw through Mabella's artifice, and leaving her question unanswered made for the door which led from the kitchen into the little front hall; whereupon Mabella deliberately placed herself in Mrs. Ranger's way, and animated by the courage which springs from consciousness of a good cause, dodged every attempt of that irate person to pass her. Mrs. Ranger endured this as long as she could, then, without more ado, she put out a strong arm and brushed Mabella aside. "Take care," she said and passed into the hall. But Sidney had had his greeting, and Vashti's calm face baffled her inquiring looks.

"I could see there had been something," she said in reporting the matter, "but what had happened I don't know."

"My sakes," said Mrs. Simpson, when Mrs. Ranger told her this, "I'm sure you *must* have been busy in the kitchen if you couldn't spare time to watch 'em meet. My soul! If Len was worth his salt for observation he'd have kep' his eyes open. But sakes! Men's that stoopid——. But with you there I thought we'd know how things was goin'——"

"Well," said Mrs. Ranger tartly, "you can thank Mabella Lansing for that. First as I was going out she ups and asks me a question. I paid no attention to that for I knew 'twas done to hinder (them Lansings is all in the same boat), and then when she seen I wasn't to be took in with that she deliberately put herself in the way, and dodged me back and forward till I had all I could do to keep from giving her a good shove."

"Well, M'bella Lansing had better look out. It's a bad thing to be set up. Pride goes before a fall. And M'bella's certainly most wonderful sure of self. But Lanty wouldn't be the first young chap to——. Of course I ain't sayin' anything, but they do say——"

Mrs. Ranger waited eagerly to see if her friend would commit herself to a definite statement. But Mrs. Simpson was much too wary for that; so Mrs. Ranger nodded her head, and pursed up her lips, and managed to convey the impression that "she could an' she would" unfold a tale.

But this was some days after Sidney met Vashti in the narrow hall of the Dole parsonage.

"I am here, Vashti," he whispered, kissing her.

"Yes, how glad I am!" she answered simply.

"Can I speak to you just a moment, dear, before I go to see him?"

"What is it?"

"Do you remember," he whispered hurriedly, "that you promised old Mr. Didymus that he should marry us? Vashti, I have waited so long. I tremble before the responsibility of the life I have chosen. Strengthen me with the fulfilment of your promise to better keep mine."

Just then Mr. Simpson came in.

"He's askin' if you be come yet," he said to Sidney. "I—wouldn't wait long before seein' him if I was you; he's sinkin'."

"I will come in at once," said Sidney. Mr. Simpson turned and re-entered the sick room.

Sidney turned to Vashti. At that moment Mrs. Ranger, flushed and a little ruffled by her combat with Mabella, entered the hall.

"How d'ye do, Mr. Martin?" she said, holding out her hand. "We'd be right glad t' see you if the time wasn't so sad."

"I am pleased to see you," said Sidney, in his gentle genial way, shaking hands with her. She looked from his face to Vashti's with an almost ridiculously eager scrutiny, but found herself baffled.

"You'd better go right in and see Mr. Didymus," she said. "He's bin askin' for you." At this juncture Mabella appeared, an adorably matronly Mabella.

"How are you, Sidney?" she asked. "Mrs. Ranger, I'm afraid your pies are burning or running over or something, I smelt them."

"Laws," said that good woman, disappearing like a shot. "Didn't you have sense enuff to go to the oving instead o' coming t' me?"

"If you want to talk," said Mabella coolly to Sidney and Vashti, "go into the sitting-room, and when she comes back I'll tell her you've gone in to see Mr. Didymus."

"You're an angel," said Sidney, and drew Vashti through the doorway just as Mrs. Ranger came back angrily.

"Them pies ain't half cooked," she said, "let alone burning!"

"Well, I'm sure I thought I smelt them," said Mabella, "and I know you didn't want to leave the pie-making for Temperance to do when she came this evening."

"If the pies had burned I'd have made others, depend on that," said Mrs. Ranger. "I guess Temp'rins Tribbey never had to do anything over after me! I s'pose he's gone in to see Mr. Didymus now?"

"We may as well go," said Mabella. "He won't be back for awhile likely."

So the two went back to the kitchen, where Lanty, after watering the roan, stood eating biscuits from the heap upon the bake board.

"Vashti," said Sidney, taking her in his arms, "say yes. You know I adore you—and—Vashti, you will——"

She looked into his eyes. For one moment a womanly hesitation prevailed in her heart. The next she questioned herself angrily: "Why wait, why delay, why not begin to lay the threads of your revenge?"

"But"—she paused and looked down. He drew her closer.

"Darling, it is the knowledge that you are really mine that I want. You surely do not think I would be exacting to you? You shall come to me when you will; say yes, dear——"

"It is so hurried—so—you *are* good," she said, with charming affectation of hesitancy.

"Send Lanty over for your father," said Sidney, "and Temperance and I will go in and ask Mr. Didymus."

"I—yes, Sidney, I will do as you wish," she said, then for one instant, abashed by the great glad light in his eyes, she let fall her face upon his breast.

"And Vashti—after—you won't keep me waiting too long."

She looked at him, arch rebuke in her eyes.

He reddened.

"There," he said, "I'm spoiling it all I know. Go, dear, and send Lanty." She moved away a step. He followed her swiftly and caught her to his breast with passion.

"Tell me, Vashti," he said, "that you love me as I love you; tell me that life together seems the only thing possible to you." She put her arms about his neck.

"I love you dearly," she said, "I could not look forward to life except with you."

With those words and with the embrace of her soft warm arms, every doubt or shadow died in Sidney's heart. He returned her embrace, too moved to speak, and left her to enter the room of the dying man.

Vashti went to the kitchen door and called her cousin.

"Lanty," she said, "will you come a moment?"

He left Mabella and came to her.

"Come outside," she said, "I want you to do something for me." Then as they got beyond Mrs. Ranger's hearing she continued: "I want you to go over and fetch father and Temperance. Sidney is bent upon being married by Mr. Didymus and—I have consented." There was a kind of agony in the regard she gave Lanty. "Will you go?" she said; her voice sounded far away to herself, and all at once it seemed to her as if she could hear the blood rushing through her veins, with a roaring as of mill-streams. And Lanty, all unconscious of this, stood smiling before her. Truly, if Vashti Lansing sinned, she also suffered.

"It's a capital idea," said Lanty heartily. "You are a lucky girl, Vashti. I'll go at once; have you told Mabella yet?"

The pent-up forces of Vashti's heart leaped almost beyond the bounds.

"Go," she said, with a strange sweet shrillness in her voice. "Go, at once."

"I will, of course, I will," said Lanty, and he suited the action to the word. He paused an instant to tell Mabella, and added: "You go and talk to Vashti, she's as nervous as you were."

Then he departed and Vashti watched him, wondering a little why she had been born to such a perverse fate. As she turned from the empty distance where he had disappeared it was to be met by Mabella's arms, and kisses, and congratulations, and exclamations. Poor Mabella! All was so well meant, and surely we would not blame her; and yet, though a creature be worthy of death, we do not like to see it tormented and baited. Vashti Lansing, with her lawless will, her arrogant self-confidence, her evil determination, was yet to be pitied that day.

The short autumnal day had drawn down to night. Lamps twinkled from every room in the parsonage. A great stillness brooded over the house.

The kitchen was filled with whispering women, groups of men lingered near the house and horses were tied here and there to the palings. The word had gone abroad that the old man who prayed for them so long was leaving them that night. There would be little sleep in Dole during its hours.

"The license has come," whispered Mabella to Temperance, and Temperance slipped out from among the women and found Nathan where he loitered by the door.

Soon they were all gathered in the sick room. Old Lansing, and Mabella, and Lanty with their baby Dorothy in his arms, and Temperance and Nathan, and another guest, unseen and silent, to whom they all did reverence, who was nearer to the old clergyman than any of them.

And in a moment the door opened, and Sidney and Vashti came softly in, both pale, both calm.

The old clergyman looked up at them lovingly. His face was the colour of ivory, and the spirit seemed to shine through its imprisoning tabernacle like a light.

In few and feeble words he married them. Then he essayed to speak a little to them, but he stumbled and faltered, and instead of saying, "You, Vashti," he said "You, Martha," and when he sought to find Sidney's name he could only say "Len."

The composure of the women gave way. Mabella buried her face in Lanty's arm and cried unrestrainedly. Tears streamed over Lanty's face also. Those words, Martha and Len, showed how lovingly, despite his stern denial of their suit, the old man had thought of his daughter and her sweetheart.

His voice wandered and failed. Sidney and Vashti knelt beside the bed.

Temperance stole forward and touching them, motioned for them to go.

As they rose the old man looked at them. A little bewilderment flickered into his eyes.

"It's not Martha and Len"—then his eyes cleared. "I am going to them and the mother." Then he looked at Sidney, "Be thou faithful unto death," he said, the solemnity of the words gaining an incalculable force from the weakness of the voice. Then he began to murmur to himself, "I have fought the good fight, I have finished my course."

Dr. Harrow and Mr. Simpson entered the room, and the others quitted it, and hardly were they gone ere the unseen guest stole out from the shadows and looked into the old man's eyes. There was neither fear nor reluctance in them, nothing but welcome, and a trust which was transcendent; and in a few moments the unseen guest folded the longing spirit of the old man in a strong embrace and bore it to where "beyond these voices there is peace."

Thus Sidney was married. Thus the mantle of the pastorate of Dole fell upon his shoulders.

CHAPTER X

For six months Sidney had been minister of Dole, and already his people adored him. Never had they heard such sweet and winning sermons; never had they realized the beauty and tenderness of the gospel, never had they gone to their church with such assurance of comfort as they did now.

As Sidney learned to know them better and better, he was enabled to comprehend more and more fully the narrow lives they led, the petty poverties which afflicted them, the sore struggle it was for most of them to make ends meet. Swayed by his great sympathy he sought in Holy Writ for all the words of comfort, peace, and promise. He read these passages to them in a voice which yearned towards them from his very heart, and then he would close the Bible and preach to them lessons of the sweetest and purest morality, illuminated by illustrations drawn from the fields they tilled, from the woods, and from the varied phenomena of natural life as it was manifested about them; his discourses came to them with a sweet and homelike sense of comfort. Dumbly and instinctively they loved their barren hills and meagre meadows with a great love, and it seemed to them that now they were being given reasons for the love which was in them.

If Sidney did not preach Christ he at least preached His word—and in His spirit, and the people to whom he preached never doubted of the chaos which was in the soul of their teacher. Their teacher who night and day kept their joys and sorrows in his heart.

Sidney was walking home through the powdery snow to the parsonage when he met Temperance; her face was set, and she was evidently in some distress of mind. One of Sidney's first pastoral duties had been to marry Temperance and Nathan. They were established in the old Lansing house, for Nathan had rented the farm. Old Mr. Lansing lived with them.

"Well, Temperance!" said Sidney. "It's an age since I've seen you; how's everyone with you?"

"Oh, well," said Temperance, "but"—looking at him shrewdly—"it don't seem to me that you are over and above well yourself."

Sidney laughed carelessly.

"Oh—I'm always well—except for the headaches, and Vashti cures them."

"Yes, I'll be bound she does," said Temperance irascibly. "You ain't got a mite of sense neither one of you; them passes and performances ain't good for you. I don't believe in 'em, and for a minister! Sakes! they say you are an angel in the village; take care you don't get to be one."

"Then you have your doubts about my being angelic?" said Sidney laughing.

Temperance coloured, but did not give way.

"Men's men," she said; "only some of them are better nor others," then she paused and grew grave and troubled again.

"You've something worrying you," said Sidney kindly; "what is it?"

"Well," said Temperance, "I don't know if I'm over anxious or not, but—have you heard anything about Lanty lately?"

"Yes, I did," admitted Sidney, "and I was terribly sorry to hear it. Do you suppose it can be true?"

"I don't want to believe it," said Temperance, two bright spots burning on her cheeks; "but—but—well—Nathan was over at Brixton to-day, and Lanty was there, and he was—not himself."

"Oh, poor Mabella!" said Sidney; "I'm so sorry. I never dreamt it could be true. What can be done?"

"Nothing—that I know of," said Temperance. "M'bella's close as wax and quite right too, but she's got a worried look; I can see through M'bella, and as for Lanty, well—it would be a pretty brave one that would speak to Lanty—he has a look!"

Sidney was in truth more distressed than he could say. That Lanty, bold, bright, honest-hearted Lanty should give way to intemperance was grievous. Sidney had always entertained a great admiration for the young countryman, who was indeed almost the antithesis of Sidney. The simplicity of his nature was very charming to this supra-sensitive man who scourged his own soul with introspective inquisition. Lanty's calm and careless acceptance of the facts of life, without question as to why and wherefore, his happy life of work with his wife and child, seemed to Sidney something to be admired as very wholesome, if not envied as being very desirable. That he should imperil this happiness seemed most tragic to Sidney.

After he parted from Temperance he walked slowly on.

It was true; Lanty had "a look." His bold eyes which had once looked so fearlessly into all the eyes they met had now changed a little. There was a kind of piteous challenge in them as of one who should say to his fellows "accuse me if you dare." Alas, over-eager denial is often an admission of guilt. The tongues had been hissing his name from house to house for long in Dole, and gradually the conviction spread that Lanty Lansing was drinking much and often—and it was true.

It was the direct result of his popularity. He had been going very often to Brixton during the past year, and there he had fallen in with a set of men who drank a great deal; the country lawyers, an old toper of a doctor, a banker and two or three idle men who spent their time in the back rooms of their friends' offices. Mixed up with this set Lanty did his drinking unseen; but, alas! the effects were very visible. But strange to say up to this time not one of the Dole worthies had seen him drunk.

It would seem that even chance was constrained to aid Mabella Lansing in the really heroic efforts she made to hide her degradation from the censorious little world about her. That she and her husband were in any sense divisible she never dreamed. Her comprehension of the unity of marriage forbade that. That Lanty could sin apart from her, or be judged apart from her, or condemned apart from her never occurred to her simple loyal mind. As for turning upon his delinquencies the search-light of her righteousness; or posing as a martyr and bespeaking the pity of her friends as so many modern wives do—well, she had none of that treachery in her. She suffered all his repentances in her own proper person and without the anæsthetic poison which sometimes numbed him to the pain of his regrets.

At this time Mabella's little child was a source of ineffable strength and solace to its mother. Its yellow head, so like Lanty's own, brightened the days he was making so dark. Mabella, grown afraid to look at the future, spent many hours in contemplating her baby. Its eyes—like bits of the blue heaven; the tiny feet whose soles were yet all uncalloused by the stones of life; the clinging hands which had as yet let fall no joy, nor grasped any thorns—these were joys unspeakable to this mother as they have been to so many. Truly "heaven lies about us in our infancy," and now and then from the celestial atmosphere about this child a warm sense of peace, a saving thrill of hope, reached out to the mother's heart. O wonderful woman heart, which, like the wholesome maple, gives forth the more sweetness the more it is pierced!

Her neighbours took up the habit of visiting her frequently. Going early and staying late, with the laudable intention of forcing themselves into a confidence denied them.

To see Lanty pass to Brixton was a signal to start to his house, there to talk to Mabella until such time as Lanty returned; and poor Mabella, all her old-fashioned wifely fidelity up in arms, talked to them bravely. They had sharp ears these mothers in Israel, but not so sharp as to outstrip Mabella's love-quickened senses.

When Lanty came back she heard his horse afar—before he came to the fork in the road even—and making some simple excuse to her visitor, she would speed out at the back door, see him, know if all was well. If his gait was unsteady and his blue eyes dazed, she would persuade him to go quietly up the back way. Happily at such times he was like wax in her hands. Then she would return to her visitor with some little lie about straying turkeys or depredating cows.

Oh, Eternal Spirit of Truth! Are not these lies writ in letters of gold for our instruction amid the most sacred precepts?

Once indeed Lanty did come into the room where Mrs. Simpson sat. His eyes were blurred; he swayed a little and asked loudly for the baby.

"I will find her," said Mabella quietly, though her heart sickened within her, and rising she led him from the room.

"Lanty, dear, you'll go upstairs and lie down?"

He looked at her white face; the truth gradually struggling in upon him; without a word he turned and crept up the back stairs like a beaten dog going to hide.

Mabella returned to the sitting-room taking her baby with her; she felt that she needed some fount of strength whilst encountering Mrs. Simpson's talk. When she entered, Mrs. Simpson greeted her with an indescribable pantomime of pursed-up lips, doleful eyes, uplifted hands and lugubrious shakes of the head. Even Mrs. Simpson dared not seek in words to break down Mabella's reticence, so baffling and forbidding was its wifely dignity.

Mabella regarded Mrs. Simpson's pantomime quietly.

"Are you not feeling well, Mrs. Simpson?" she asked. "Are you in pain?"

Mrs. Simpson arrested her pantomime with a jerk, and sitting very erect, quivering with righteous wrath and excitement over the exclusive information she possessed, she said:

"I'm real well—I am. I only thought—but I guess I'm keeping you; p'raps you've got other things to do. Isn't Lanty needin' you?"

"No," said Mabella, "Lanty is not needing me. What made you think that? And I hope you'll stay to tea. I've just put the kettle forward."

"No—I can't stay," said Mrs. Simpson. "I only came to visit for a while and I've stayed and stayed." Mrs. Simpson had at the moment but one desire on earth, which was to spread the news of Lanty's fall.

"I sort o' promised to visit Mrs. Ranger this week. I've visited a long spell with you now. I guess I'll be going on. My! How like her father that young one do grow!"

"Yes, doesn't she?" said Mabella, and the gladness in her voice was unfeigned.

Mrs. Simpson took the goose quill out of her apron band, in which her knitting needle rested, and measured the stocking she was knitting with her second finger.

"Well!" she said, "I declare I've done a full half finger sence I been settin' here! This is my visitin' knittin'. I hain't done a loop in this stockin' but what's been done in the neighbours'. I cast it on up to Vashti's. My soul! I never can come to callin' her nothing but Vashti, if she be the minister's wife! I cast it on up there, and the preacher he was real took up with the three colours of yarn being used at once. He sez, sez he: 'Why, Mrs. Simpson, you're all three fates in one: you have the three threads in your own hands.' Then he said to Vashti, 'That would be fittin' work for you, Vashti.' Well, I knowed Vashti could never manoover them three threads at once, but I didn't say nothin', bein' as I thought he was took up with the stockin' and wanted Vashti to make him some. Then he told about some woman named Penellepper that was great on knittin'. The only girl I ever knowed by that name was Penellepper Shinar, and she certingly was a great knitter; she used to knit herself open-work white-thread stockings. Well, she came to a fine end with her vanities! I wonder if 'twas her Mr. Martin meant? Folks did say she was living gay in Boston, though 'twas said too that she went fur west somewheres and school-teached. Suz! It would be queer if 'twas her Mr. Martin meant!"

"Mr. Martin gets all those stories out of old books, in learned tongues," said Mabella simply. "When he stayed at the farm he used to tell us all sorts of stories."

"Women in books is mostly bad 'uns," said Mrs. Simpson, by this time arrayed in the old *crêpe* bonnet which had been bought as mourning for Len, and which she now wore as second best. "That holds good even to the Bible and the newspapers. And as for a preacher mixing himself up with them, I don't hold with it. But being that they're mostly dead it don't matter so much, and judging from all accounts they was good riddance when they died."

What a requiem over the "dear dead women" to whom so many songs have been sung!

"How that scented geranium grows! It beats all," said Mrs. Simpson, as Mabella escorted her to the garden gate. For anyone to have let a visitor depart alone from the doorstep would have been a scandal in Dole.

"Won't you have a slip?" said Mabella, setting down Dorothy and bending over the plant. "It's apple scented; Lanty bought it off a pedler's waggon over in Brixton in the spring; it has grown wonderfully."

She broke off a branch, ran for a bit of paper, put a little ball of earth round the stem, wrapped it up and gave it to Mrs. Simpson.

"Well, it's real generous of you to break it, Mabella; but you know the proverb, 'A shared loaf lasts long.'"

"Yes, it's true I'm sure," said Mabella.

She accompanied Mrs. Simpson to the gate and held up the baby to wave good-bye.

And Mrs. Simpson sped down the road with the fleetness of foot which betokens the news bringer.

She turned at the fork in the road and looked back at the square house against its background of trees. Mabella was still at the gate with the yellow-headed baby.

"Well," said Mrs. Simpson to herself, "them Lansings is certainly most tormented proud! Sich pretences! And would I stay to tea! My! I wonder Mabella Lansing can look a body in the face. Gracious! She must think we're a set of dumbheads, if she thinks every soul in Dole can't see how things is goin' with Lanty. It's the drinkin' uncle coming out clear in him that's sure."

Mrs. Simpson arrived at her friend's house in ample time for tea, and under the stimulus of excitement made an excellent repast.

Without criticism upon the Dole people it must be admitted that a scandal in their midst, such as this, had much the same exhilaration about it for them that a camp meeting had.

Mrs. Simpson and Mrs. Ranger talked over all the ins and outs of the Lansing family history. It was all equally well known to each, but after all, it is an absorbing and amusing thing to rake over well-hoed ground.

Public opinion had long since been pronounced upon the events which these two worthy women cited, not only that, but the grist of diverse opinions had been winnowed by the winds of time till only the grain of public decision was left.

So that when Mrs. Simpson expressed her opinion emphatically in regard to any point, she knew Mrs. Ranger would agree with her, and knowing every link in the chain of events, knew exactly what would be suggested to the other's memory by her own remark.

But it is a great mistake to think these conversations devoid of mental stimulus. It required great adroitness to prevent the other person from seizing upon the most dramatic situations and making them hers.

Then, too, though this was an unholy thing, there was always the odd chance that an opinion, differing from that pigeon-holed in the Dole memory as correct, might be advanced. In this case it was one's bounden duty to strive by analogy, illustration, and rhetoric, to bring the sinner back to the fold of the majority.

Nor must it be supposed that history handed down thus, crystallized. The light and shadows were for ever shifting, and when any new incident occurred the other cogent incidents in the chain were instantly magnified and dilated upon, and for the time being stood forward boldly in the foreground of the pedigree under consideration, remaining the salient points until such time as some new event shed lustre upon another set of incidents.

In view of the sensation of the moment, the "drunken uncle" loomed like an ominous spectre across the long vistas of the Lansing genealogy. For the moment he was regarded as the direct progenitor of all the Lansings, although he had died unmarried fifty years before Lanty's birth.

Mrs. Simpson added another half finger to her fateful stocking, with its triune thread, ere she quitted Mrs. Ranger's that night.

"Well, I declare," she said, as she stood on the step in the greyness of the falling night. "I declare! I most forgot the slip Mabella gave me. It's on the bed where my bunnit was," she added to little Jimmy Ranger, who went in search of it. "It's real rare that geranium is, apple scented—smell," breaking off a leaf, pinching it, and holding it under Mrs. Ranger's nose. "Come up as soon as you can," she added, descending the two steps.

"Yes," said Mrs. Ranger, "we're going to Brixton for the blankets that have been spun of last year's wool, next week, and p'raps we'll drop in on the way home."

"Do," said Mrs. Simpson, "and you kin stay supper and visit a spell; our cider'll be made by then. Len's been over to the cooper about the mill this week. But if you should hear anything in the meantime, jest put on your bunnit and come acrost the fields neighbourly."

"Yes, I will," said Mrs. Ranger; "I guess things is comin' to a head; I wouldn't be surprised any day——"

There was a long pause.

"Nor Me," said Mrs. Simpson emphatically. "Good night."

"Good night. It gets dark real soon now."

"Yes, there's quite a tang to the air to-night. It'll be frost in no time."

"Well," soliloquized Mrs. Simpson, as she betook herself home, "Liz Ranger thinks just the same's I do; that's evident. My sakes! How Mabella Lansing can go through with it is more'n I can figure."

"It's terrible!" said Mrs. Ranger, going back leisurely to the house. "It's downright terrible. I guess Lanty went on awful to-day. Mrs. Simpson is jest full of it, but sakes! I should think she'd kind of talk low of drinkin' and sich, remembering her own Len. He was a rip, Len Simpson was, if ever there was one! But that don't seem to be a bridle on Gert Simpson's tongue. It's enough to bring a jidgment on her, the way she talks. I wonder how Temp'rins Peck 'll like Lanty's goin's on?"

These reflections of Mrs. Ranger's upon Mrs. Simpson were no doubt edifying, but certainly she had carried on the conversation with quite as great a gusto as Mrs. Simpson. And if she had not enjoyed it as much it was only because Mrs. Simpson, being a redoubtable conversationalist, had filched the finest morsels of the retrospective talk for herself, it was therefore probably more a sense of wounded *amour propre* than genuine condemnation of Mrs. Simpson which led her to criticize the latter's conversational methods.

Mrs. Ranger had an uneasy and unsatisfactory idea that she had merely given Mrs. Simpson her cues.

Mabella made strong coffee that night for supper instead of tea. She dressed Dorothy in the beribboned dress that Sidney had sent from Boston. She talked cheerily and brightly to her husband. She rose from her place and came round with his cup and put it beside him, letting one hand fall with a passing but loving touch upon his shoulder as she did so. But she did not look at his face once during all the time of supper. She dreaded to see the crown of shame upon the brow of her king. For herein again Mabella showed the steadfastness of her adherence to her husband. She suffered because he suffered. It was not the fear of the scandal that would arise, it was not the thought of her own probable future which stung her to the heart, although these thoughts were both bitter as wormwood.

It was the knowledge that Lanty, her Lanty, who was her guide, her everything, was ashamed. It was the harm he was doing himself that she deplored, not the reflection of his behaviour upon herself.

How many the women who proclaim their own patience and their husbands' shortcomings upon the housetops think of this? Not long since a certain woman, bediamonded and prosperous, was demanding sympathy from her dear half-dozen friends, recounting to them the derelictions of her husband. "There's only one comfort," she said; "after every break he makes, he always gives me a handsome present. That's always something." Yet we wonder that there are cynics!

There was no word spoken between Lanty and Mabella in reference to the afternoon. But that night in the darkness Lanty suddenly drew her into his arms.

She laid her cheek against his; both faces were wet with tears.

There was poignant apology made and free, full loving pardon given all in that instant.

And Mabella wept out her pain on his breast.

But the shame and bitterness and self-contempt ate into Lanty's heart like a venomous canker....

All this had been in the late autumn, just after the death of old Mr. Didymus, and now it was spring, and all through the winter Mabella had suffered, and hoped, and prayed, and despaired, and now it had come to this that Nathan had seen Lanty intoxicated in Brixton!

Sidney went back to the parsonage sorely troubled at heart. Vashti stood in the doorway.

Her beauty struck him freshly and vividly. It was his whim that she should dress in rich and beautiful stuffs, and Vashti was quite willing to subscribe to it. Dole groaned in spirit at the spectacle of its minister's wife in such worldly garb as she wore, but Dole would have borne much at Sidney's hands.

To-day she was clothed in a softly draped house-gown of Persian colouring, bound by a great cord girdle about her waist; it fell in long classic lines to her feet. Vashti's face had gained in majesty and strength since her marriage. She was thinner, but that, instead of making against her beauty, raised it to a higher plane. There was a certain luxuriousness in her temperament which made her rejoice in the beautiful things with which Sidney surrounded her. She felt instinctively that she gained in forcefulness and in individuality from her setting. And, indeed, she fitted in well amid the beautiful pictures and hangings with which Sidney had adorned the enlarged parsonage. She had always seemed too stately, too queenly, for her commonplace calicoes and cashmeres. Her mien and stature had made her surroundings seem poor and inadequate. But in this gem of a house she shone like a jewel fitly set. Sidney had had his own way about the primary arrangements, and had installed a strong working woman in the kitchen with Sally, the ex-native of Blueberry Alley, as her under-study.

Vashti was perfectly content with this, and, whilst she knew all Dole was whispering about her, held upon her way undisturbed. She had developed, to Sidney's intense joy, a very decided taste in the matter of books. Her mind was precisely of the calibre to take on a quick and brilliant polish. She read assiduously, and her perceptions were wonderfully acute.

Her beginnings in literary appreciation were not those of a weakling. Her mental powers were of such order that from the first she assimilated and digested the strong, rich food of the English classics.

She delighted in verse or prose which depicted the conflict of passion and will, of circumstances and human determination. Alas, her education only made her more determined to gain her purpose, more contemptuous of the obstacles which opposed her.

And yet, if her purpose had not been of the most steadfast, she might well have been discouraged.

Lanty and Mabella seemed so securely happy. Vashti was, however, gaining an ascendency over her husband which almost puzzled herself. She had no comprehension whatever of the nature of the power by which she was enabled to cause a deep mesmeric sleep to fall upon him. Nor did she understand in the least how gradually but surely she was disintegrating his will. When his headaches came on now half a dozen gestures of her waving hands were sufficient to induce the hypnoses which brought him forgetfulness. Ignorant of the potency of suggestion she often stood watching him whilst he slept, feeling within her the striving of her dominant will, as of an imprisoned spirit striving to burst the confining bars.

"Come into the study," said Sidney, as he reached her side. "I have some very bad news."

"My father?" she said.

"No, Lanty." She blanched to the tint of the powdery snow. Together they went to the study, and he told her.

Her breath came quickly.

Was the longed-for opportunity to be given into her hands at last?

With all her mental activity she could not yet guess how Lanty's decadence might yield her the opportunity she craved.

But the position of affairs had seemed so barren of hope for her that any change seemed to make revenge more near.

So the evil in her leaped and strove upward like a flame given fresh fuel and freer air.

CHAPTER XI

The fragrant pink arbutus had replaced the snow-wreaths upon the hillsides, the downy whorls of the first fern fronds were pushing through the dark-brown leaves, the fragile hepaticas had opened their sweet eyes wide, when one morning Sidney took the sloping path which led up the hill overlooking Dole.

His face was pale and drawn, his grey eyes half distraught, his slender, nervous hands clinched as if to hold fast to some strand of hope, some last remnant of courage, some crumb of consolation for that moment when his soul, utterly bereft, should cry aloud in desolation.

Sidney Martin preached to his people sweet and wholesome sermons, instinct with the hopefulness and charity of one who believes that, "all things work together for good," and that "the mute beyond is just," but in his own soul was chaos.

Always sensible of his personal responsibility towards his fellows, he had now become almost morbid upon the subject.

The old workman had known Sidney better than Sidney had known himself, and his prophecies were being fulfilled.

Happy as Sidney was in his husbandhood, yet the possession of Vashti was not a narcotic strong enough to stupefy his keen spiritual nature.

Every Sunday before he entered the pulpit he endured a Gethsemane; every time he quitted it he sought the faces of his people yearningly, pitiably, eager to be assured that his words had comforted them.

He spent all his time thinking of and for them, and he had won closer to their hearts than he guessed. They gave him confidences which had been withheld from their fellows for years, and thus let in to the closed chambers in their humble lives, he was able to justify himself to Vashti for the very lenient way in which he looked upon their lapses. He sometimes wondered that their common experiences of poverty and effort did not make them more considerate in their judgments upon each other. But they found in him always a merciful judge. He visited their homes, he knew their hopes and fears, he appreciated the pathos of their narrow ambitions, at which a less great-hearted man might have laughed.

He went into the little school-house frequently, and strove in simple words to awaken the children to the beauty about them, to the possibilities of life. He had great hopes of the children. Already he had singled out several whom he thought might make scholars. He promised himself that they should be given the opportunity.

He had been going to the school that morning when a little incident occurred which awakened all his most poignant doubts of himself, and the righteousness of his ministry.

Passing by the school-playground, he had seen some evil words chalked up in a school-boy hand upon the board fence. It was like a blow in the face to Sidney—so eager to instil the doctrines of sweetness and light into these children. Why, O why had that boyish hand traced the symbols to form that evil idea? It was as if a clear spring should suddenly cast up mud instead of water.

Sidney effaced the words, but turned away from the school. The whole morning was poisoned for him. Poor Sidney! Doubtless he was suprasensitive, and yet—why had not the boy chosen some sweet and beautiful words to write upon that sunny spring morning? Surely they would have been more in keeping with the whole world as the boy's eyes saw it?

We may smile at Sidney as he agonizes alone upon the hill, but it was by such vigils as these that he won so close to the heart of the God in whom he had no belief.

Sidney wandered about in the woods upon the hillside till gradually some little of the peace of the day entered into his spirit. He gathered a bunch of arbutus to take home to Vashti. He encountered no one upon the return journey but Mr. Simpson, who "passed the time of day" with the minister, as he said afterwards, and then proceeded to try to draw him out regarding Lanty. It was very easy for Sidney to parry old Mr. Simpson's queries, but they made him very uneasy nevertheless.

Vashti whitened as Sidney related the circumstances to her.

Could there be anything new? she wondered. Sidney had one of his intense headaches, and, after the mid-day meal, Vashti proposed to give him ease from it by putting him into a sleep.

"You are my good angel, Vashti," he said, catching her fingers as she made the first pass across his forehead, and kissing them one by one. She looked down at him, for he lay upon the green leather couch in the study, and smiled almost tenderly. His continual sweetness of temper, his unselfishness, his thoughtfulness, and, above all, his great adoration for her had touched her greatly since their marriage. She was too keen an observer, too clever a woman, not to recognize that this man was head and shoulders above the men she had known. She had moments when she was enraged against herself for loving Lanty instead of her husband, but yet her heart never wavered in its allegiance to her yellow-haired cousin. There was something in his magnificent physique, his superabundant energy, his almost arrogant virility, which appealed to her. Beneath that calm, pale face of hers were strong passions, sleeping, but stirring in their sleep at the voice which did not call them.

Sidney, or Sidney's welfare, would never weigh with her a featherweight if balanced against a chance of winning Lanty from her cousin, or of revenging herself upon them both, yet there were times when she wished that it had been any other man than Sidney who was bound to her.

"It is you who are good," she said. "The village people think you are a saint."

"Vashti," said Sidney, wistfully. "Do you think I do them good?"

"Indeed, yes," said Vashti, "just think how they turn out to church. It's something wonderful."

Sidney's eyes lighted up with delight of her praise.

"Oh, Vashti!" he said, "I am so glad. I often wonder if you are satisfied with my work. You know it was you who ordained me to the priesthood."

A slow colour stole into her cheeks. She waved her hands soothingly above his brow, then posing two fingers upon his temples where the pain was, said gently but imperatively, "*Sleep, sleep*," and almost immediately, with her name upon his lips, he closed his eyes and fell into a deep slumber.

She leaned back in her chair and looked about the room, so manifestly the sanctum of a man of taste. The bookshelves which extended round and round the room to the height of a man's shoulder, were filled with books uniformly bound in dark green leather.

This was a miracle in Dole, and Sally was wont to dilate upon the astonishing circumstance, and marvel that Mr. Martin could find the one he wanted among so many all alike. The mere fact of the titles being different did not appeal to Sally.

Above the bookshelves, against a soft harmonious background, were beautiful etchings from the paintings Sidney loved. Millet's peasants, Burne-Jones' beautiful women, Meissonier's cavaliers, Rossetti's "Beata Beatrix." Upon the top of the bookshelves were two exquisite marbles, the winged Victory of Samothrace, and the Venus de Milo, and one bronze—the famous wing-footed Mercury, slender, lithe, and seeming ever to sweep on with the messages of the gods.

Vashti sat long there, then she remembered that it was the day of the sewing circle. The meeting was at the house of Mrs. Winder that day.

Vashti rose and left the room; she put on her hat, paused to look at herself in her glass, and smiled to think of how the women would whisper, when her back was turned, about her Boston gown and her modish hat.

Vashti rather liked to amaze her fellow-women. With all her strength of mind there was much femininity about her, and when it came to prodding up other women she was an adept.

As she passed the open study door she paused and looked in where her husband lay, sunk in the unconsciousness of a hypnotic sleep. For a moment she had a great desire to awaken him, but still softened by unwonted tenderness, she refrained from doing so. Vashti liked not only to parade her Boston finery before the sewing circle, but also her husband.

After all, being the minister's wife in Dole had charms.

"If I had only told him to come for me," she said regretfully. "I wish he would, at five o'clock. I've a mind to wake him up and ask him." She hesitated. The light slanted in across Sidney's face, its pallor shone out startlingly.

She turned away and ere long was nearing Mrs. Winder's. She walked slowly up the path to the front door. Sidney often forgot that it was one of the preacher's privileges to do this, but Vashti always remembered what was fitting; besides, she knew the window of the sitting-room commanded the little path, and she thought the sewing circle might just as well be edified by her progress from the gate as not.

"My! Vashti is most terrible cherked up in her dress," said Mrs. Ranger to Mrs. Winder.

"Yes, that gownd must have cost a lot, but they say 'tis by the preacher's wish."

"Who said that?" asked Mrs. Simpson.

"Well," said Mrs. Ranger volubly, "I heard that too; it was Sally, up at the preacher's, that told young Mary Shinar, and Mary Shinar told Tom, and Tom had it over to our Ab at Brixton a week come Saturday, that the preacher draws the patternings for Vashti's gownds, and colours them himself, and measures Vashti with a tape line, and sends the hull thing off to somewheres in Bosting, and Sally up at the preacher's says that when they come from Bosting the sleeves and the waist is all filled full of silk paper to hold 'em in shape, and that it's like a body in a coffing when the lid is taken off, and—yes, my turkeys has been laying for a week now," concluded Mrs. Ranger with an abrupt change of subject and tone, for

Vashti at that moment entered the room. Now Vashti herself had ere now switched off her conversation to a side track, and when she heard Mrs. Ranger answering a question which had not been asked, she smiled in a manner to make even Mrs. Ranger uncomfortable.

Vashti had hardly taken her place before Temperance entered, and presently the twenty or thirty women were busy with their needles upon the somewhat formless garments which are supposed to conduce to the salvation of the heathen, and whilst their needles were busy their tongues kept pace.

There were many things of importance to be discussed, the health of Vashti's father (who had had another stroke), the setting of hens, the finding of turkeys' nests, house cleaning and garden making—the springtime in the country is always a busy time—and above and beyond all these things there was a most exciting subject, the downfall of a certain Ann Serrup; of this the matrons whispered together.

"Has Mr. Martin been over yet?" asked Mrs. Winder of Vashti, after trying in several indirect ways to find out.

"No," said Vashti, "I don't think he has heard of it. I didn't tell him and I don't think anyone has."

"If you take my advice," said Temperance, making her needle whistle through the cotton. "If you take my advice you'll keep the preacher away from that mess. He's that soft-hearted that he's liable to be taken in—besides, it's more likely a woman's help she needs. Laws, I ofting think of Ann, all alone. Why don't you go yourself, Vashti?"

"I have thought of it for a couple of months," said Vashti. "It's nearly a year old now, isn't it?"

"Yes," said Mrs. Winder, proceeding to give data. "But sakes! Why couldn't she stay over Brixton way without coming into our parish with her brat?"

"They have souls," said Vashti, suddenly drawing the mantle of the preacher's wife about her.

"Well, one of 'em shouldn't have," said Mrs. Ranger irately. "Sakes, I don't know what girls is coming to!"

"I expect she didn't have much chance," said Temperance deprecatingly.

"That's no excuse for sin," said Vashti austerely.

Temperance sniffed audibly. The clock struck five, and a footstep sounded upon the porch of the backdoor.

"Run see who that is," said Mrs. Winder to Jimmy.

The women held their needles suspended midway in the stitch, and Sidney's voice came cheerily from the kitchen.

"Why, lands sake! It's Mr. Martin and by the kitching too!" said Mrs. Winder bustling forward to welcome him.

He entered gracefully, greeting them all in his gentle genial way which seemed to bring him so close to their hearts; but his eyes sought out Vashti where she sat half anticipative—half dreaming of the words he would say. Somehow it seemed to her that she was taking part in a scene which had been rehearsed long since and which grew slowly into her recollection. Sidney would say—she thought the words and Sidney's voice seemed the audible echo of the phrase, "You wanted me to come at five," he said; "I just woke up in time; it was fortunate I did not forget. Are you going over to see your father?"

"Yes," said Vashti, rising mechanically, a strange mingling of awe and exultation, not unmixed with fear, at her heart.

"You will excuse my wife if she is lazy to-day, Mrs. Winder," said Sidney laughing, "but I hope you won't follow her bad example and leave off before the six o'clock bell; we must have full time in the sewing-class!"

There was a general smile at this mild wit. Ministers' jokes are always highly appreciated.

"What a beautiful view you get from this window," said Sidney, looking out across to the hill. Mrs. Winder saw her opportunity and took it.

"Yes," she said, "but you get a terrible fine view from the window in the front room—just step in, if you'll take the trouble." So saying Mrs. Winder threw open the door of the sacred front room, revealing all its glories to Sidney's gaze, and preceding him with a great assumption of unconsciousness, she rolled up the paper blind and pointed out of the window.

Sidney looked, and saw almost opposite him a new frame barn whose pine walls showed glaringly and somewhat oppressively in the sun.

"The new barn 'ill be done in two weeks," said Mrs Winder as Sidney turned away; "you see it lengthways from here."

"It looks very well," said Sidney kindly. Then he bade them all good-bye and departed with Vashti, who was silently marvelling. This was the first inkling Vashti had of the force of "suggestion."

Meanwhile the tongues buzzed in the company they had left. The women were conversationally inclined; excitement is a great stimulant to the flow of ideas, and certainly this meeting of the sewing-circle had had its sensation. Mrs. Winder's boldness in inveigling the preacher in to see the glories of the front room had been appreciated at its full worth. Not one of these dames but had cherished a secret longing to show off her front room to Sidney—but so far he knew only the mundane comfortableness of the "setting-rooms."

Mrs. Winder had scored largely that day.

And the meeting was not over.

Mrs. Ranger had been irritated that afternoon in various ways. Vashti's smile when she entered had made Mrs. Ranger uncomfortable.

"Although," as she said to Mrs. Winder, "what could she expect? My sakes! I don't care if she did hear me! It's all gospel truth and what can she expect, being the preacher's wife, but to be talked about?"

What indeed?

Then, too, Mrs. Ranger felt Mrs. Winder had indulged in reprehensibly sharp practice in regard to the front room—and—but it is needless to enumerate the different irritations which, combined, made Mrs. Ranger venomous. She felt she must ease the pressure upon her patience by giving someone's character a thorough overhauling; so with a side look at Temperance, and a tightening of her meagre mouth, she began to speak of Lanty.

Now in Dole, if any subject was brought up which hurt or pained you, you were expected to look indifferent, make no reply, and strive by keeping a calm front to deny the honour of putting on the shoe when it fitted.

The Spartan boy's heroism has often been outdone by women who smiled and smiled whilst venomous tongues seared their hearts. So Mrs. Ranger began boldly, as one does who fires from under cover at an unarmed foe.

But Temperance had been so long one of the Lansing family that she had assimilated a little of their "unexpectedness," and as Mrs. Ranger continued her remarks, egged on by acquiescing nods from the older women, there began to gather upon the brow of Temperance a deep black cloud.

Mrs. Ranger paused in her harangue to gather breath for her peroration, when suddenly the thread of talk was plucked from her ready lips by the strident voice of Temperance, who, rising to her feet, and gathering her sewing together as she spoke, proceeded to deliver herself of an opinion upon the charity of the women about her. In whatever particular that opinion erred, it certainly merited praise for its frankness. After Temperance had indulged in a few pungent generalizations she narrowed her remarks to Mrs. Ranger's case. Never in all the annals of Dole had any woman received such a "setting out" from the tongue of another as Mrs. Ranger received that day from Temperance. Temperance spoke with a knowledge of her subject which gave play to all the eloquence she was capable of; she discussed and disposed of Mrs. Ranger's forbears even to the third generation, and when she allowed herself finally to speak of Mrs. Ranger in person, she expressed herself with a freedom and decision which could only have been the result of settled opinion.

"As for your tongue, Mrs. Ranger, to my mind, it's a deal like a snake's tail—it will keep on moving after the rest of you is dead."

With which remark Temperance departed from the sewing circle which had metaphorically squared itself to resist the swift onslaught of her invectives; she gathered her skirts about her as she passed through the room, with the air of one fain to avoid contamination, and stepping forth as one who shakes the dust from off her prunella shoes as a testimony against those she is leaving, she took the road home. Temperance's mouth was very grim, and a hectic spot burned the sallowness of her cheeks, but she said to herself as she strode off briskly:

"Well—I 'spose it's onchristian, but it's a mighty relief t' have told that Mrs. Ranger just once what I think of her—but oh! pore Lanty and pore, pore M'bella! To think it should come about like this!"

And the red spots upon her cheeks were extinguished by bitter tears.

The sewing circle broke up in confusion; one could only hear a chorus of "Well—I declare!" "It beats all!" "Did you ever!" as the ladies bundled their work together—each eager to get home to spread the news and to discuss the matter with her husband.

And that night in the starlight Mabella waited at the little gate listening for the hoof beats of Lanty's horse from one side, and the cry of little Dorothy from the house behind her.

And when Lanty came—alas! What "God's glowworms" in the sky revealed, we shall not say.

But we will echo the words of Temperance—"Pore Lanty—pore, pore M'bella!"

CHAPTER XII

The Ann Serrup of whom the sewing circle had whispered, was one of those melancholy scapegoats found, alas! in nearly every rural community, and lost in cities among myriads of her kind. She had lived in the Brixton parish all her life, but had lately come with her shame to a little house within the precincts of Dole. Left at thirteen the only sister among four drunken brothers much older than herself, the only gospel preached at—not to—her had been the terrorism of consequences. Like all false gospels this one had proved a broken reed—and not only broken but empoisoned. The unfathered child of this poor girl had been born about a year prior to her appearance in Dole.

Mabella's heart went out to the forlorn creature, and a few days after the memorable meeting at Mrs. Winder's she set forth to visit her, leaving Dorothy in charge of Temperance. It was a calm, sweet season. The shadow of white clouds lay upon the earth, and as Mabella walked along the country roads the chrism of the gentle day seemed to be laid upon her aching heart. For a space, in consideration of the needs of the poor creature to whom she was going, Mabella forgot the shadows which dogged her own steps.

She was going on a little absent-mindedly, when at a sudden turn in the road she came upon Vashti, who had paused and was standing looking, great-eyed, across the fields to where the sun smote the windows of Lanty's house.

"Well, Mabella," she said, taking the initiative in the conversation as became the "preacher's wife." "Where are you going?"

"I'm going to see Ann Serrup," said Mabella. "I've wished to do so for some time—how plainly you can see our house from here."

"Yes—how's Lanty?"

"He's very well—haven't you seen him lately?—he looks splendid."

"I didn't mean his looks," said Vashti with emphasis.

"Well, one's looks are generally the sign of how one feels," said Mabella bravely, although she winced beneath Vashti's regard. "And Vashti, Dorothy can speak, she——"

Vashti broke in with the inconsiderateness of a childless woman.

"Do you know anything about Ann Serrup? Is she penitent?"

"I—I don't know," said Mabella hesitatingly (she had heard most unpromising accounts of Ann's state of mind, "Fair rampageous," Temperance had said), "she has suffered a great deal."

"She has sinned a great deal," said Vashti sententiously.

They walked on almost in silence, and ere long stood before the low-lying, desolate dwelling.

A girl came to the open door as they drew near—poorly but neatly clad, and with tightly rolled hair. A girl in years—a woman in experience. A child stood tottering beside her.

"Come in," she said to them before they had time to speak, "come in and set down."

She picked up the child, and unceremoniously tucking him under one arm, set two chairs side by side; then put the baby down and stood as one before her accusers. Her brows were a little sullen; her mouth irresolute. Her expression discontented and peevish, as of one weary of uncomprehended rebuke. The baby clutched her dress, and eyed the visitors placidly, quite unaware that his presence was disgraceful.

Mabella looked at the little figure standing tottering upon its uncertain legs; the little dress was so grotesquely ill-made; the sleeves were little square sacks; the skirt was as wide at the neck as at the hem. She thought of her well-clad Dorothy and her heart went out to the desolate pair.

The mother, tired of Vashti's cold, condemnatory scrutiny, began to shift uneasily from one foot to the other.

"What's your baby's name?" asked Mabella, her sympathies urging her to take precedence of the preacher's wife.

"Reuben," said Ann.

"Reuben *what?*" demanded Vashti in sepulchral tones.

"Jest Reuben—Reuben was my father's name"—then with fretful irritation—"jest Reuben."

"Is your child *deformed?*" asked Vashti suddenly, eyeing with disfavour the little chest and shoulders where the ill-made frock stuck out so pitifully.

"Deformed!" cried Ann, the pure mother in her aroused; "there ain't a better-shaped baby in Dole than my Reub."

She sat down upon the floor, and, it seemed to Mabella, with two movements, unclothed the child, and holding him out cried indignantly—

"Look at him, Missus Martin, look at him! and if you know what a baby's like when you see one you'll know he's jest perfect—ain't he, Missus Lansing? Ain't he? You know, don't you?"

Vashti glared in fixed disapproval at the baby, who regarded her not at all, but after a leisurely and contemplative survey of himself began to investigate the marvels of his feet, becoming as thoroughly absorbed in the mysteries of his own toes as we older infants do in our theories. "He's a beautiful baby—I'm sure you are very proud of him," said Mabella kindly. Then her gaze rested upon the two poor garments which had formed all the baby's costume. Tears filled her eyes as she saw the scrap of red woollen edging sewn clumsily upon the little yellow cotton shirt.

"I'm afraid you are not used to sewing much," she said, "it was the clothes which spoiled the baby."

Ann, who, unstable as water, never remained in the same mood for ten minutes together, began to cry softly, rocking back and forth sometimes.

"Oh, I wisht I was dead! I do. I never was learned nothing. 'Scuse me if I spoke up to you, Missus Martin, but I'm that ignerent! And you the preacher's lady too! My! I dunno how I came t' be so bad. I guess I'm jest real condemned bad; but I haint had no chance, I haint; never a mother, not so much as a grandma. Nothing but a tormented old aunt. And brothers! Lord! I'm sick of brotherses and men. I jest can't abear the sight of a man, and I'm that ignerent. Lord! I can't make clo'es for Reub, now he is here." Then vehemently—"I am jest dead sick of men."

"But, think," said Mabella soothingly, "when Reuben is a man he'll look after you and take care of you."

"Yes—I s'pose he will," said Ann, drying her eyes; then, with a sudden change of mood, she began smiling bravely. "Say—he's that knowin'! You wouldn't believe it; if I'm agoin' out in a hurry I give him sometimes an old sugar rag, but he knows the difference, right smart he does, and he jest won't touch it if 'taint new filled; and"—with a touch of awe as at a more subtle phenomenon—"he yawned like a big person when he was two days old."

"Why, so did my baby," said Mabella in utter astonishment that another baby had done anything so extraordinary.

"Are you coming, Mabella?" said Vashti austerely from the doorway.

Direst disapproval darkened her countenance. Ann's mutable face clouded at the words.

"Yes, I'm coming," said Mabella hastily to Vashti, then she turned to Ann. "I will send you some patterns to cut his dress by," she said. "It's very hard at first; Temperance helped me; I'll mark all the pieces so that you'll know how to place them," then she went close to Ann and put a trembling hand upon her arm.

"Ann," she said, "promise that you'll never do anything wicked again—promise you'll never make your baby ashamed of you."

"No, I won't; I've had enough of all that—you'll be sure to send a pattering with a yoke?" inquired Ann eagerly.

Poor Ann! Her one virtue of neatness was for the moment degraded to a vice; she so thoroughly slighted the spirit of Mabella's speech. But Mabella, out of the depths of her motherly experience, pardoned this.

"Yes, I will send the nicest patterns I have," she said.

"Soon?"

"Soon—and Ann—you'll come to church next Sunday?"

Ann began to whimper.

"Oh, I hate t' be a poppyshow! and all the girls do stare so, and——"

"Ann," said Mabella pleadingly, "you'll come?"

"Yes, I'll come, Missus Lansing, being as you want me to," then another swift change of mood overtook the poor, variable creature.

"They kin stare if they want to! I could tell things! Some of 'em ain't no better nor me if all was known. I'll jest come to spite 'em out. You see—I'll be there."

"I shall be so glad," said Mabella gently, having the rare wisdom to ignore side issues. "I'll see you, then."

"Oh, Lor'," said Ann, whimpering again, "ye won't want to see me when other folkses are around, and I s'pose you've got a white dress and blue ribbings for church, or red bows, like as not. Lor'! Lor'! what 'tis to be born lucky. 'Better lucky nor rich,' I've heard said ofting and ofting, and it's true, dreadful true. I never had no luck; neither had mother; she never could cook anything without burning it, and when she dyed 'twas allus streaky! I've heard Aunt Ann say that ofting and ofting; he *is* a fine baby, isn't he?" she broke off abruptly.

"Yes, indeed," said Mabella heartily. "Good-bye, Ann," and stooping she kissed the girl and went out and down the path. Ann stood gazing after her.

"She kissed me," she said dully, then in an echo-like voice repeated "kissed me."

The old clock ticked loudly, the kettle sang on the fire, the baby fell over with a soft thud upon the floor. Ann sat down beside him, and clasping him to her breast cried bitterly to herself, and as has been often the case, the mother's sobs lullabyed the child to a soft and peaceful sleep. She rose, with the art which comes with even unblessed motherhood, without waking the child, and laid him down gently.

"I know she won't send a pattering with a yoke," she said in the tone of one who warns herself against hoping too much.

Meanwhile Mabella sped after Vashti; she overtook her in about a mile.

"Goodness, Vashti," said Mabella; "I'm sure you need not have hurried so! I'm all out of breath catching up."

"Well, I couldn't stand it any longer," said Vashti.

"Stand what?" demanded Mabella, a little irritated by Vashti's ponderosity of manner.

"That exhibition," said Vashti with a gesture to the forlorn house, which somehow looked pitiably naked and unsheltered. "It was disgusting! To go about petting people like that is putting a premium upon vice."

Mabella laughed.

"You dear old Vashti," she said, "you said that as if you had been the preacher himself—what the world could I say to her? standing there with that poor child." A sudden break interrupted her speech. "Oh, Vashti," she said, "isn't it terrible? Think of that baby; what a difference between it and Dorothy! And so poor—so very poor; without even a name; Vashti—you're a lot cleverer than me; you don't think, do you, that they will be judged alike? You don't think there will be one rule for all? There will be allowances made, won't there?"

"I wonder at you, Mabella," said Vashti, "putting yourself in a state over that girl and her brat! It's easy seeing you've precious little to trouble you or you'd never carry on about Ann Serrup; a bad lot the Serrups are, root and branch; bad they are and bad they'll be. The Ethiopian don't change his spots! and as for crying and carrying on about her! take care, Mabella, that you are not sent something to cry for—take care." The last ominous words uttered in Vashti's full rich voice made Mabella tremble. Ah—she knew and *Vashti knew* how great cause she already had to weep.

"How can you talk to me like that?" she said to Vashti passionately; "how can you? One would think you would be glad to see me in trouble. If it's any satisfaction to you to know it I may as well tell you that——" Mabella arrested her speech with crimson cheeks. What had she been about to do? To betray Lanty for the sake of stinging Vashti into shame.

"Dear me," said Vashti coolly; "you are growing very uncertain, Mabella!"

"Yes, I know," stammered Mabella. "Forgive me, Vashti."

"Oh! It doesn't matter about my forgiveness," said Vashti; "but it's a pity to let yourself get into that excitable state."

They were near the spot where their ways parted.

Mabella looked at Vashti, a half inclination to confide in her cousin came to her. It would be such a help to have a confidant, but her wifely allegiance rose to forbid any confidences regarding her husband's lapses; she must bear the burden alone. A lump tightened her throat as she closed her lips resolutely. These little victories seem small but they are costly.

"Good-bye, Mabella," said Vashti; "come over and have tea with us soon."

"I'll come over after dinner and stay awhile with you," said Mabella, "but I won't stay to supper."

"Oh, why?" said Vashti. "Lanty can come in on his way home from Brixton; if he turns off at the cross road he can come straight up Winder's lane to the parsonage. He's often at Brixton, isn't he?"

"Yes," said Mabella, once more calm in her *rôle* of defender. "Yes, but I'll come over some day after dinner; Lanty likes supper at home. He's often tired after being in Brixton. I'll bring Dorothy and come over soon for a little visit."

"Well, you might as well come all of you for supper," said Vashti; and somehow by a subtle intonation of the voice she conveyed to Mabella the fact that her unconsciousness was only feigned.

As Mabella went towards home the lump in her throat dissolved in tears; she allowed herself the rare luxury of self-pity for a little space, then with the instinctive feeling that she must not give footing to such weakness she pulled herself together, and went forward where Lanty waited at the gate.

When Vashti turned away from Mabella to take the little path to the parsonage, her heart also was wrung by regret and pain; she had made Mabella feel, but how gladly she would have exchanged her empty heartache for the honour of suffering for Lanty's misdeeds. Lanty Lansing was very handsome, very winning, with that masterful tenderness and tender tyranny which women love; but it is doubtful if he (or many other men) deserved the love which these two women lavished upon him. And it must be said for Vashti that whatever her faults were, she loved her cousin well and constantly. His excesses rent her very heart; if she saw in them a hope of vengeance upon Mabella she yet deplored them sincerely. The hate which was growing in her heart against Mabella was intensified a thousand-fold by the thought that she did not, in some way, drag Lanty back from the pit. Had she been his wife she would have saved him in spite of himself. The thought that the village was sneering and whispering about her idol made her eyes venomous, and in this mood she entered the house. Sidney was waiting for her and suddenly there swept across the woman's soul a terrible sense of the relentless Destiny which she was working out. As in a mirror she saw herself, not the free and imperious creature she had imagined, but a serf, shackled hand and foot, so that her feet trod the devious path prepared for them from time immemorial, and her hands wrought painfully at a fabric whose fashion and design were fixed by other power than her own.

And Sidney, with his pale spiritual face, his unearthly exalted eyes, his eager-winged soul, was bound to her side. His footsteps were constrained to hers, only it seemed that whereas the path was chosen for her, his way was simply outlined by her will; she remembered the strange incident which had taken her away from the sewing circle. Again she experienced the thrill, half of fear, half of mad unreasoning triumph, which had held her very heart in suspense when Sidney had said, "You wished me to come at five." Could it be that whilst his mind was passive, whilst he slept the sleep her waving hands induced, whilst his faculties were seemingly numbed by the artificial slumber, could it be that he could yet grasp her desires and awake to fulfil them? The simplest knowledge of hypnotic suggestion would at once have given her incalculable command over Sidney. As it was, she could only grope forward in the darkness of half fearful and hesitating ignorance. In her advance to the knowledge that Sidney, whilst in this sleep, was amenable to suggestion (although she did not phrase it thus) she had skipped one step which would have given her the key to the whole; she had seen that he would carry out, whilst awake, a wish of hers expressed whilst he slept. She did not know that he would have been a mere automaton in her hands whilst he was in the hypnotic sleep, but she told herself that she must measure and ascertain exactly the control she had over her husband; thus nearly every day she cast the spell of deep slumber upon him and gradually, little by little, she discovered the potency of suggestion.

It must be said that Sidney was entirely acquiescent to her will. The old weird fables of people hypnotised against their wills have long since been relegated to the limbo of forgotten and discredited myths; and while it is certainly true that each hypnosis leaves the subject more susceptible to hypnotic influence, it is utter rubbish to think that influence can be acquired arbitrarily without the concurrence of the subject. But Sidney had given himself up to the subtle delight of these dreamless slumbers as the hasheesh-eater delivers himself to the intoxication of his drugged dreams.

Sidney's mind was torn by perpetual self-questionings; not about his own personal salvation, but about his responsibility towards the people of Dole. The more he studied the Bible the more deeply he was impressed by the marvellous beauty of the Christ story. Never surely had man realized more keenly than Sidney did the ineffable pathos and self-sacrifice of the Carpenter of Galilee. Often as he passed the little carpenter-shop where Nathan Peck came twice a week, he entered and stood watching Nathan planing the boards, and as the long wooden ribbons curled off before the steel, and the odour of the wood came to his nostrils, quick with that aroma of the forest which obtains even at the core of the oak, there surged about Sidney's heart all the emotions of yearning and hope, and sorrow and despair which long, long ago had lifted That *Other* from a worker in wood to be a Saviour of Souls; and he went forth from the little carpenter-shop as one who has partaken of a sacrament. And often he stood upon the little hill above Dole, his eyes full of tears, remembering that immortal, irrepressible outburst of yearning, "O, Jerusalem, Jerusalem, how often would I have gathered thy children together, as a hen doth gather her brood under her wings and *ye would not*!"—the poignancy of this plaint wrung Sidney's very soul. And how sweet it seemed to Sidney to steal away from all these questions and questionings, to fall asleep with Vashti's eyes looking, as it seemed to him, deep down into his very soul, seeing the turmoil there and easing it with the balm of her confidence and strength—to awaken with the knowledge that there was something Vashti wished done, something he could do. Thus, whereas the occasions of Sidney's acute headaches had been formerly the only opportunity Vashti had had of experimenting with this new and wonderful force which she so dimly understood, now it was a daily occurrence for Sidney to cast himself down upon the green leather couch and seek from Vashti the gift of sleep.

Thus, gradually, surely, Vashti won an ascendency over this man which made him in every sense her tool. Happily she did not know the full extent of her power. But if knowledge is power, certainly power brings knowledge, and thus it was that ere long Vashti was turning over in her mind the different ways and manners in which she could apply this power of hers. Thus equipped with her own unfaltering resolution and having the energy of a second person at her command, Vashti brooded over her plans.

The night after Mabella's visit to Ann Serrup, Lanty was at home, and seated before the open door, was coaxing plaintive melodies from out the old fiddle, which having been regarded as a godless and profane instrument for several generations in his father's family, had at last fallen upon happy days and into appreciative hands, for Lanty Lansing could bring music out of any instrument, although, of course, he had never been taught a note. The old fiddle under Lanty's curving bow whispered and yearned and moaned and pleaded—the dusk fell and still he played on and on—till Mabella, having put Dorothy to bed, came out to sit upon the doorstep before his chair, resting her head against his knee. The fiddle was put down. For a little the two sat in silence. Afterwards the scene came back to them and helped them when they had sore need.

"Lanty," said Mabella, "will you do something for me to-morrow?"

"What is it?"

"Oh, Lanty!" reproachfully.

"Of course I'll do it; but I can't, can I, unless I have some slight idea."

"Well, you are right there," she said; "I thought you were going to object! Well, you know Ann Serrup?"

"I know her, yes; a precious bad lot she is too!" Lanty's face clouded.

"Lanty, dear, wasn't that just a man's word? She's a woman, you know, and Lanty, I've been to see her, and it's all so forlorn; and she's so—so—oh, Lanty! And Vashti was there and she asked if her baby was *deformed*, fancy that! And it was the poor little scraps of clothes which made the child look queer. But it was the sort of queerness which makes you cry, and Lanty, I said I would send her some patterns, and you'll take them over to-morrow morning, won't you?"

"But, girlie," he began; just then Dorothy gave a sleepy cry.

Lanty and Mabella rose as by one impulse and went into the twilight of the room where the child's cot was. But their baby slept serenely and smiled as she slept.

"The angels are whispering to her," said Mabella. The old sweet mother fable which exists in all lands.

"Yes," said Lanty. A tremor shaking his heart as he wondered why this heaven of wife and child was his.

As they passed into the other room they saw the child's clothes upon a chair in a soft little heap like a nest; and all at once there rushed over Mabella's tender heart all the misery of that other mother, and before Lanty knew it, Mabella was in his arms crying as if her heart would break.

"Oh, Lanty, Lanty," she sobbed; "think of poor Ann Serrup! When her baby cries in the night who goes with her to look after it?"

"There, there," said Lanty, searching distractedly for soothing arguments; "don't, Mabella, don't; I'll take the traps over first thing in the morning." And presently Mabella was comforted, and peace rested like a dove upon the roof-tree.

So early next morning Lanty departed with the parcel. In due time he arrived before the little house. The house door stood open—humbly eager to be entered. Early as it was Ann was up, and came to the door looking neat and tidy. She took the parcel with the undisguised eagerness of a child. Lanty turned away, letting his horse walk down the lane-like road. He was not much given to theorizing; a good woman was a good one, a bad one a bad one in his estimation, but this morning he found himself puzzling uneasily over the whys and wherefores. It is an old, old puzzle, and like the conundrum of Eternity, has baffled all generations, since the patriarch of Uz set forth that one vessel is created to honour and another to dishonour. So Lanty found no solution, and was tightening his reins to lift his horse into a gallop when he heard someone calling, and turning, saw Ann speeding in pursuit. She reached him somewhat blown and decidedly incoherent as to speech.

"She has sent the yoke pattering, and a white apron and heaps of things! There ain't nothing but real lady in Mis' Lansing! Sakes! I wisht the preacher's wife could see Reub now! I'll take him to church next Sunday, and if he squalls I can't help it And here—take this and keep it—and don't let him harm me, will you? And I never meant no harm to you personal, but he was for ever pestering me, and he said he was coming over early this morning for 'em, and for me to sign 'em; but Lor'! I didn't have no ink—and don't tell Mis' Lansing, she's a lovely lady, and I didn't mean no harm, and he said there wouldn't be no law business, because you'd give me heaps of money, 'cause being as you drank, people would believe anything of you; and Lor'! hear that baby! Mind you don't tell Mis' Lansing"—with which Ann turned and fled back to console her child. Lanty, much mystified, opened the thin packet of papers. An instant's scrutiny sent him into a blind mad rage, which made him curse aloud in a way not good to hear.

For before him, writ fairly forth in black and white, was a horrible and utterly baseless accusation, purporting to be sworn to by Ann Serrup, and witnessed by Hemans, the machine agent of Brixton.

The witness had signed his name prematurely before Ann, and had written faintly in pencil "Sign here" for her benefit and guidance.

Lanty gathered the import of the papers and put them securely in his pocket.

He was just opposite a thicket of wild plums, shooting up through them was a slim and lithe young hickory. Lanty flung the roan's bridle over a fence fork, cut the young hickory, and remounting went on his way. Only he turned away from Dole, and proceeded slowly towards Brixton, and presently, just as he entered the shadow of Ab Ranger's wood by the roadside, he saw a blaze-faced sorrel appear round the bend and he rejoiced, for he knew that his enemy was given into his hand....

Hemans was sorely bruised when Lanty flung him from him with a final blow and a final curse. He tossed aside the short fragment of the young hickory which remained in his grasp.

Lanty's fury had lent him strength, and he had well-nigh fulfilled the promise made in the first generosity of rage to thrash Hemans "within an inch of his life."

"And now," said Lanty, addressing Hemans with a few words unavailable for quotation. "And now, open your lips if you dare! If you so much as mention my name, I'll cram the words and your teeth down your throat. Remember, too, that I have something in my pocket which would send you where you'd have less chance to prowl. And, mind you, don't try to take it out of Ann Serrup. If you do I'll finish your business once for all. Paugh! Vermin like you should be knocked on the head out of hand. If I stay I'll begin on you again——." Lanty swung himself up on the roan.

"Don't make any mistake as to my intentions," he called over his shoulder. "I've given you one warning, but you won't get two."

Hemans lay groaning upon the ground, and just about that time Ann, having dressed her baby in the white pinafore Mabella had sent, came to her door, and, leaning against it, looked forth at the morning.

She thought of Hemans and the papers.

"The fat's in the fire now," she said, smiling inanely, divided between vague curiosity over the outcome and gratification over the baby's appearance in its new finery.

Lanty had given Hemans salutary punishment, but his heart sickened within him.

He knew what a leech Hemans would have proved if he had once got a hold upon him; and if he had refused to be blackmailed——?

Lanty knew well with what insidious, untraceable persistency a scandal springs and grows and spreads in the country. He knew how hard it is to kill, how difficult to locate, like trying to catch mist in one's hands. He had heard often that wicked proverb which says, "Where there's smoke there must be some fire." A man has often self-possession to extricate himself from a danger, the retrospect of which makes him nearly die of fear. And so it was with Lanty, as there grew upon him the sense of what a calamity might have overtaken him.

How Mabella might have been tortured by this horrible falsehood. Mabella, his wife, who blushed still like the girl that she was! It was a very tender greeting he gave his wife and child when he reached home, and he made Mabella very happy by his account of Ann's delight over her gift. And then he strode off to his fields, and all day long he remembered two things—that Mabella's charity to the poor disgraced girl had already brought its blessing back to the giver, and that one phrase of Ann's, "being as you drank."

Lanty had never realized fully before what he was doing. But his eyes opened. He could look forward to the future, but the thought of the past gave him a sense of helplessness which made his heart ache.

With every honest effort of his hands that day he registered a vow. The peril he had escaped had opened his eyes to the other dangers which threatened the heaven, which he had thought he possessed so securely, of wife and child.

The real purification of Lanty's life from the sporadic sin which had beset him took place that day as he worked in his fields, but his friends and neighbours always thought the change dated from another day a few weeks later.

For although we have learned our lesson well, yet Destiny, like a careful schoolmaster, takes us by the hand, and leading us over sharp flints and through thorny thickets, revises the teachings of our sufferings.

CHAPTER XIII

Three days after Lanty's interview with Hemans, Mabella paid a visit to Vashti.

Sally, grown in stature if not in grace, promptly carried off Dorothy, and the two cousins sat down opposite each other in the dainty room which served as a sitting-room and drawing-room in the Dole parsonage.

There was a great contrast between the two women; despite the beauty and hauteur of Vashti's face there was a shadow of ineffable sadness upon it. Life was none too sweet upon her lips.

The seed sown in barren Mullein meadow had brought forth a harvest of bitter herbs—wormwood and rue, smartweed and nettles.

Shadowing her eyes was the vague, ever-present unrest of those who do battle with spectres of the mind; there is no expression more pitiful, because it speaks of unending warfare. But upon her brow there shone the majesty of an unconquered will; she had not been bent beneath the knee of man's authority, nor ground into the mire by poverty's iron heel, nor bowed beneath the burden of physical pain.

She was in some strange way suggestive of the absolute entity of the individual.

Human ties and relationships seemed, when considered in connection with her, no more than the fragments of the wild vine, which, having striven to bind down the branches of the oak, has been torn from its roots by the merciless vigour of the branch to which it clung, and left to wither without sustenance.

Now and then against the background of The Times there stands forth one figure sublimely alone, superimposed upon the fabric of his generation in splendid isolation—a triumphant, individualized *ego*.

It is almost impossible to study and comprehend these individuals in their relations to others, the sweep of impulse and energy, the imperious flood of passion, the tumultuous tide of his which animates their being and stimulates their actions is so different from the sluggish, well-regulated stream whose current controls their contemporaries.

They *must* be regarded as individuals; adown the vista of the world's perspective we see them, splendid; but eternally alone in the centre of the stage, brilliant and brief, like the passing of a meteor coming from chaos, going—alas! almost inevitably—to tragedy; leaving a luminous trail to which trembling shades creep forth to light feeble lamps of imitation, by which to trace the footsteps of the Great Unknown.

But we never understand these people, who, great in their good or evil, baffle us always—defying the scalpel which would fain anatomize them—now and then, as by revelation, we catch a glimpse of their purpose, a hint of their significance, but when we would fix the impression it eludes us as the living sunshine mocks at the palette of the painter, and spends itself royally upon the roof-trees of peasants, when we would wish to fix it for ever in unfading pigments and hang it upon the walls of kings' palaces.

In her degree Vashti Lansing was one of these baffling ones.

Compared with her cousin Mabella, she was like a beautiful impressionistic picture beside a carefully designed mosaic.

The one compound of daring and imagination, gorgeous in colour, replete with possibilities if barren of achievement, offending against every canon, yet suggesting a higher cult than the criticism which condemns it; the other typical of the most severe and elaborated convention, executed in narrow limits, yet charming by its delicacy and stability, an exponent of the most formal design, yet winning admiration for the conscientiousness with which its somewhat meagre possibilities have been materialized.

Yet Mabella Lansing's face was eloquent.

It was composite of all the pure elements of womanhood—the womanhood which loves and bears and suffers but does not soar. In her eyes was the soft fire of conjugal and maternal love. With the tender, near-sighted gaze of the home-maker, her eyes were bent upon the simple joys and petty pains of every-day life.

Upon her countenance there shone a tender joyousness, veiled but not extinguished by a certain piteous apprehension; indeed, there was much of appeal in Mabella's face, and bravery too—the bravery of the good soldier who faces death because of others' quarrels and faults.

But above all it was the face of the Mother.

Surely no one would dispute the fact that motherhood is the crowning glory of woman, the great holy miracle of mankind; but while it is impious to deny this, it is unreasonable and absurd to say that for all women it is the highest good.

There are different degrees of holiness; even the angels differ one from the other in glory; why, then, should the same crown be thought to fit all women?

The golden diadem may be more precious, but shall we deny royalty to the crown of wild olive or to the laurel wreath?

The mother is the pole-star of the race, but there are other stars which light up the dark places; why should their lonely radiance be scoffed at?

Women such as Mabella Lansing are the few chosen out of the many called.

There was in her that intuitive and exquisite motherliness which all the ethics on earth cannot produce. A simple and not brilliant country girl, she yet had a sense of responsibility in regard to her child which elucidated to her all the problems of heredity.

It is probable that she was a trifle too much impressed with her importance as a mother, that she had rather too much contempt for childless women, but that is an attitude which is universal enough to demand forgiveness—it seems to come with the mother's milk—yet it is an unlovely thing, and whilst bowing the head in honest admiration of every mother, rich or poor, honest or shamed, one would wish to whisper sometimes to them that there are other vocations not lacking in potentialities for good.

"What a lovely house you have, Vashti!" said Mabella, irrepressible admiration in her voice, a hint of housewifely envy in her eyes.

"Yes, it is very comfortable," said Vashti, with a perfectly unaffected air of having lived in such rooms all her days.

"Comfortable!" echoed Mabella; then remembering her one treasure which outweighed all these things, she added, a little priggishly: "it's a good thing there are no babies here to pull things about."

Vashti smiled in quiet amusement.

"What's the news in the village?" she asked. "You know a minister's wife never hears anything."

Mabella brightened. Good little Mabella had a healthful interest in the social polity of the world in which she lived, and Vashti's disdain of the village gossip had sometimes been a trial to her. Vashti usually treated "news" with an indifference which was discouragingly repressive, but to-day she seemed distinctly amiable, and Mabella proceeded to improve the opportunity.

"Well," she said, "the village is just simply all stirred up about Temperance's quarrel with Mrs. Ranger. I always knew Temperance couldn't abide Mrs. Ranger, but I never thought she'd give way and say things, but they do say that the way Temperance talked was just something awful. I wasn't there; it was at the sewing circle, and for the life of me I can't find out what started it, but, anyhow, Temperance gave Mrs. Ranger a regular setting out. I asked Temperance about it, but the old dear was as cross as two sticks and wouldn't tell me a thing. So I suppose it was something about Nathan. Young Ab Ranger has got three cross-bar gates making at Nathan's shop, and they've been done these three days, and he has never gone for them; he's fixing up the place at a great rate. I suppose you know about him and Minty Smilie? Mrs. Smilie's going about saying Ab isn't good enough for Minty; and they say Mrs. Ranger is just worked up about it. I wouldn't be at all surprised if matters came to a head one of these days, and Ab and Minty just went over to Brixton and came back married—" suddenly Mabella arrested her speech and a more earnest expression sweetened her mouth. "Vashti," she resumed, "there is something I wanted to ask you. Ann Serrup sent me word that she was coming to church next Sunday, and I want you to speak to Sidney and get him to preach one of his lovely helpful sermons for her. I'm sure he will if you ask him. Something to brace her up and comfort her, and, Vashti—I'm awfully sorry for her." Mabella paused, rather breathlessly and a little red; "one never knew exactly where one was" with Vashti, as Temperance was fond of saying.

For a fleeting instant during Mabella's little recital Vashti's eyes had contracted in almost feline fashion, but she replied very suavely:

"I'll tell Sidney, but, well—you know I never interfere in the slightest with his sermons."

"Oh, no," said Mabella with really excessive promptitude; "Oh, no, you wouldn't dare to do that."

"Of course not," said Vashti with so much of acquiescence in her voice that it was almost mocking.

"I know how men think of these things," continued Mabella with the calm front of one thoroughly acquainted with the world and its ways. "But Sidney is different; he is so good, so gentle, and he seems to know just how one feels"—a reminiscent tone came to Mabella's voice, she recalled various hours when she had needed comfort sorely and had found it in the gracious promises Sidney held out to his listeners. "It is a great comfort to me," she went on; "lately it has seemed to me as if he just held up the thoughts of my own heart and showed me where I was strong and where lay my weakness." Mabella arrested herself with an uncomfortable knowledge that Vashti was smiling, but when Vashti spoke a silky gentleness made her voice suave.

"I will tell Sidney what you say, and no doubt he will preach with a special thought of you and Ann Serrup."

"Well, I'm glad I spoke of it," said Mabella; "I wasn't sure how you'd take it."

Vashti continued to smile serenely, as one who recognizes and understands cause for uncertainty. Her gaze was attracted to the window.

"Look!" she said suddenly.

Passing in plain view of the window was a most extraordinary figure. A creature with a face blacker than any Ethiopian, surmounted by a shock of fair hair—this individual was further adorned by the skirt of a bright blue dress, which, being made for a grown-up woman, dragged a foot or so on the ground behind; about the neck was a pink silk tie, showing signs of contact with the black, which was evidently not "fast"; above her head she held a parasol bordered with white cotton lace—thus caparisoned Sally paced it forth for the amusement of little Dorothy, who tottered upon her legs by reason of the violence of her laughter. Surrounding the pair, and joining apparently in the amusement, were the two dachshund puppies (Sidney's latest importation to Dole), the collie, who followed with the sneaking expression of one who enjoys a *risqué* joke (and yet he could not forbear biting surreptitiously at the dragging flounces as they passed), and little Jim Shinar, who followed in a trance-like state of wide-eyed fascination. He lived nearer to the parsonage than any other child, and between the evil fascination which Sally exerted over him and the dread of finding himself within the gates of a man "who spoke out loud in church," Jim's life was oppressed with continual resistance to temptation, but he had frequent falls from grace, for Sally could do more things with her mouth and eyes than eat and see, indeed her capabilities in the line of facial expression were never exhausted, and there was a weirdness about her grimaces which fascinated older children than poor round-faced little Jim.

Sally peacocked it up and down before her admiring satellites, until suddenly there rang through the parsonage a vigorous expression uttered in a rich brogue, and at the same instant a large, red-faced woman rushed out of the kitchen door and appeared round the corner of the house.

Sally arrested her parade, paused, showed an inclination to flee, paused again, then with a gibe for which she dived back into her Blueberry Ally vocabulary, fled from the irate "work-lady," who had unwittingly furnished forth the fine feathers in which Sally was strutting. Mary promptly gave chase, and that too with an agility which her bulk belied. The area of the hunting ground was not very great being bounded by the prim palings of the little garden, but no landscape gardener ever made more of his space than did Sally. She doubled and turned and twisted, and eluded Mary's grasp by a hand-breadth, as she darted under her outstretched arms, but Sally was very unwise, for she used her breath in taunts and gibes, whilst Mary pursued the dishonoured flounces of her Sunday gown in a silence which was the more ominous because of her wonted volubility.

Sally was getting slightly winded, and was wishing she could get the gate open and give Mary a straightaway lead, but she had her doubts of the gate, sometimes it opened and sometimes it didn't. Sally knew if it was obstinate that her fate was sealed; she was casting about for another means of escape when her adherents began to take a share in the proceedings. First, little Jim Shinar, standing rooted to the spot, saw the chase descending upon him; Sally dodged him, but Mary was too close behind and too eager for her prey to change her route quickly, so she charged into him and went over like a shot. Jim gave a howl, and Mary gathered herself up, and, breaking silence for the first time, ordered him home in a way not fit for ears polite, and then resumed the chase; but the dachshunds, seeing their playmate little Jim in the thick of it, concluded that there might be some fun in it for them also, and promptly precipitated themselves upon Mary in a way which impeded her progress so much that Sally was able to make the gate and get it half open before Mary shook herself free, but when she did she came like a whirlwind towards the gate, cheered on by the collie, whose excitement had at last slipped the collar and vented itself in sharp barks. Sally whisked through the gate, but Mary was at her heels. Sally felt the breath of the open, and knew if she escaped Mary's first sprint that she was safe. So with a derisive taunt she sprang forward, jubilant, but alas, in the excitement of the crisis Sally let go her hold of the long skirt, which immediately fell about her heels, and in an instant the chase was ended, for Mary, panting, blown, and enraged beyond expression, was on Sally in a second, and fell with her as the long skirt laid her low—the dachshunds arrived a little later, and the collie, seduced by their evil example, threw decorum to the winds, and seizing an end of the bright flounce where it fluttered under the angry clutch of Mary, he tugged at it with might and main, and this was the scene which greeted Sidney, as, returning from his walk, he approached his own gate.

He had met a herald of the war in the person of little Jim Shinar, who was fleeing home as fast as his sturdy legs would carry him, crying at the same time from pure bewilderment.

A word and a small coin healed all little Jim's hurts, and Sidney proceeded, wondering what had frightened the child, whom he was used to seeing about the kitchen or in Sally's wake when she went errands.

Now, as was recorded afterwards in Dole, Sidney conducted himself under these trying circumstances with a seeming forgetfulness of his ministerial dignity which was altogether inexplicable, for, instead of immediately putting the offenders to open shame, he laughed, and even slapped his leg (so rumour said, though this was doubted), and called to the dachshunds, who were amusing themselves demolishing Mary's coiffure, in a way which savoured more of encouragement than rebuke.

It is hard to live up to "what is expected of us," and for once the Dole preacher was disappointing—but nevertheless, his presence brought the peace which he should have commanded. For Sally's unregenerate soul owned one reverence, one love—for her master she would have cut off her right hand. To have him see her thus! There was a violent upheaval in the struggling mass, then Sally was free of it and speeding towards the house at a rate which suggested that her former efforts had not been her best. Mary gathered herself up, and seeing Sidney, by this time outwardly grave, standing looking at her, she too made for the house, and Sidney was left still very stupefied, gazing upon the two dachshunds, which, suddenly finding themselves deprived of amusement, fell upon each other with a good will which proved them fresh in the field.

Sidney entered the house where Mabella and Vashti waited laughing.

Sidney was very pleased to see his wife's face irradiate with girlish laughter. She had been so grave and quiet of late that his loving heart had ached over it. Was she not happy, this beautiful wife of his?

She had a far keener appreciation of the real humour of the situation than had Mabella, and when her husband entered her eyes danced a welcome. He was enthralled by the sight, and was more than glad to give Mary the price of two dresses to mend her flounces and her temper. Nor did he rebuke Sally too severely for the unauthorized loan she had levied upon Mary's wardrobe. He knew Sally had been sufficiently punished by his appearance. Mabella had rescued Dorothy at the first alarm, and the child had looked upon the whole proceeding as an amiable effort on Mary's part to amuse her.

Shortly after Sidney's arrival Mabella departed, having enjoyed her visit greatly, and Lanty and she spent an hour that evening listening to Dorothy, as, with lisping baby tongue and inadequate vocabulary, she endeavoured to describe how Sally had blackened her face with blacklead to amuse her.

That night Sidney sat alone in his study; his shuttered window was open, and, between the slats, the moths and tiny flying creatures of the night came flitting in. Soon his student lamp was nimbused by a circle of fluttering wings. Now and then an unusually loud hum distracted his attention from the loose-paged manuscripts before him, and he laid them down to rescue some moth, which, allured too near the light, had come within dangerous proximity to the flame.

These poor, half-scorched creatures he sent fluttering forth into the night again, yet, in spite of this, several lay dead upon the green baize below the student lamp; others walked busily about in the circle of light cast by the lamp-shade upon the table, and presently he put aside all pretence of work and watched them with curious kindly eyes.

His heart, that great tender heart which was for ever bleeding for others, whilst its own grievous wound was all unhealed, went out even to these aimless creatures of a day.

Surely some leaven of the divine Eternal Pity wrought in the clay of this man's humanity, making it quick with a higher life than that breathed by his nostrils.

"Not a moth with vain desire
Is shrivelled in a fruitless fire,
Or but subserves another's gain,"

he said to himself, and then before his watching eyes there seemed to be mimicked forth all the brave-hearted struggle of humanity towards the light, which, alas! too often scorched and blasted those nearest to it. Well, was it better, he wondered, to have endured and known the full radiance for an instant, even if the moment after the wings were folded for ever, or was it wiser to be content upon the dimmer plane as those little creatures were who ran about the table-top instead of striving upward to the light? But happily, as he looked at these latter ones, his attention was diverted from the more painful problem, as his eyes, always delicately sensitive to the beauty of little things, dwelt with delight upon the exquisite, fragile little creatures.

How marvellously their delicate wings were poised and proportioned! Some had the texture of velvet and some the sheen of satin; and nature, out of sheer extravagance, had touched them with gold and powdered them with silver. And did ever lord or lady bear a plume so daintily poised as those little creatures bore their delicate *antennæ*? And presently a white creature fluttered in from the bosom of the darkness, a large albino moth with a body covered with white fur and two fern-like *antennæ*; white as a snow-flake it rested upon the green baize.

Just then Vashti entered, coming up to his table in her stately fashion.

"How foolish you are to sit with your window open," she said. "Don't you know that the light attracts all those insects?"

Sidney had risen when his wife entered the room. He was almost courtly in his politeness to her. But it was so natural for him to be courteous that all little formalities were graceful as he observed them.

As he rose he knocked down a book. He stooped to pick it up; as he straightened himself he saw Vashti's hand upraised to strike the white moth.

"Oh, Vashti! don't! don't!" he cried, irrepressible pain in his voice; but the blow had fallen.

The moth fluttered about dazedly, trying to escape the shadow of the upraised hand; there was a powdery white mark on the green baize table-top where the first blow had fallen upon it, maiming it without killing it outright.

Sidney's face grew pale as death.

"Oh, Vashti! Vashti!" he cried again. "Do not kill it, there is so much room in the world."

He gathered the half-crushed creature, which would never fly again, into a tender hollowed palm, and, opening the shutter, put it forth to die in the darkness from whence it had been drawn by the glimmer of his lamp.

Alas! alas! how many wounded and maimed have been cast forth to die in the darkness from out which their aspirations had drawn them to receive their death wounds. Sidney came back to his table, a sick pain at his heart.

Presently Vashti put her arms about his head, and drawing it back upon her breast, placed her cool finger-tips upon his eyes.

He accepted the mute apology with swift responsive tenderness. And as she held him thus the woman's weakness, latent even in her, forced itself to the surface for a moment.

"You suffer for every little thing," she said. "I can only feel when my very soul is torn."

He felt two tears fall upon his face; he drew her towards him; she sank beside his chair upon her knees, and he pressed her head against his breast, and she submitted to the caress and rested upon him in a sort of weary content, as one who pauses upon a hard journey; he put down his face till it leaned upon her hair, and thus, so near together that heart beat against heart, so far apart that the cry of the one soul died and was lost ere it reached the other, they remained for long, whilst before them the silver lamp and its white flame grew dimmer and dimmer, as its light was obscured by the shimmering veil of tiny creatures who danced about it.

Oh, piteous allegory! Can it indeed be that by our very efforts to find Truth we hide its radiance from others?

CHAPTER XIV

There are certain flowers which, when placed with other blossoms, choke and stifle and wither them by some evil emanation so subtle that it cannot be analyzed. The heliotrope is one of the flowers which murder other blooms. As with flowers so with spirits. Which of us that is at all sensitive to psychic influences but has felt at one time or another the devitalizing influence of certain personalities, and one can readily imagine how continuous, how fatal such an influence would be, when the eyes were so blinded by love that they could neither perceive the evil plainly nor guess its genesis at all. And sometimes thinking of these things, one wonders if the old, weird tales of vampires and wehr-wolves are not cunning allegories instead of meaningless myths, invented by men who, searching the subtleties of soul and spirit, had discerned this thing, but living in times when it was not wise to prate too familiarly of the invisible, had been fain to cloak their discovery in a garb less mystic.

But if the strife be wrapped in mystery the effect upon the subjective spirit is very visible.

Many of the Dole people eyed their pastor anxiously as he arose to address them the next Sunday, for he was very dear to them. Dole was not prone to let its affections go out to strangers. Life was very pinched and stinted in Dole, and it would seem almost as if their loves were meagre as their lives; at their repasts there was rarely much more than would go round, and perhaps they remembered better the injunction against giving the children's meat to the dogs, than they did the command to love thy neighbour as thyself. The great luxury of the poor—loving—they did not half enjoy, but bounded their affections as they did their fields.

Between Dole and strangers there was usually an insurmountable barrier of mutual incomprehension. It was, indeed, difficult to find the combination which opened the Dole heart, but Sidney had done it.

He was a very tender pastor to his people; whatever doubts, whatever questionings, whatever fears troubled and tormented his own soul, he permitted none of them to disturb the peace of the doctrine he preached. These people striving with irresponsive barren acres, and bending wearily above hopeless furrows, were told how they might lighten the labours both of themselves and others, and promised places of green pastures and running brooks. The gates of their visionary celestial city were flung wide to them, and in the windows of the heavenly mansion cheering lamps were lit.

Was this false doctrine? Perhaps. Protestants are fond of saying with a sneer that Catholicism is a very "comfortable religion." The implication would seem to be that a religion is not to be chosen because of its consolations. Therefore, it is perhaps regrettable that Sidney's preachment to Dole was so pronouncedly a message of "sweetness and light."

His hearers loved him, and looked upon his unministerial ways with a tolerance which surprised themselves; often, as he passed upon these long, seemingly aimless, walks which Dole could not comprehend, a hard-wrought man would pause in his work, straighten himself and look after him wistfully even as the eyes of the fishermen followed the Galilean, or a weary woman would stand in her doorway until such time as he drew near, and then, with some little excuse upon her lips, arrest his steps for a moment, to turn away comforted by the benediction of his mere presence.

Nor was Sidney insensible of, or irresponsive to this output of affection. He felt the full force of it, and returned to them their full measure heaped up and running over. And for a time the comfort of the mutual feeling helped to sustain his spirit, fainting beneath the burden of morbid introspection, and sapped by the ignorantly exercised power of his wife, for, not understanding the influence she wielded, Vashti used it rashly. Suggestion was superimposed upon suggestion until the centre of his mental gravity was all but lost, and in his walks he often paused bewildered at the upspringing of certain things within his mind, grasping at the elusive traces of his vanishing individuality.

The hour is past when these things might be scoffed at; the old legends have given place to scientific data more marvellous than the myths they discredit. The law has recognized the verity of these things, and justice has vindicated its decision with the extreme fiat of death. Alas, the justice of men is for those who kill the body; it cannot reach those who murder the mind.

The church was unusually crowded when Sidney arose. It had been hinted abroad that Ann Serrup was to be there, and Dole stirred with pleasurable anticipation, for Ann Serrup was an unregenerate individual so far as religion was concerned.

It was related of her that once at a revival meeting in Brixton, when the fiery revivalist of that place, Mr. Hackles, approached her, asking in sepulchral tones where she expected to go when she died, Ann replied, unmoved, that she would go to where they put her, a response calculated, in the mind of Mrs. Ranger, to bring a "judgment onto her."

The Rev. Hackles denounced her as a vessel of wrath and designated her as chaff ready to be cast into the fire, but Ann sat dreamily through it all, and, as Lanty related afterwards, "never turned a hair." And this was when she bore no other shame than the stigma of being a Serrup, and therefore predestined to evil, and now she was coming to Dole church. What would their gentle pastor say?

It was a sweet summer day. Mabella and little Dorothy sat by a window, and the yellow sunshine lingered about the two yellow heads, and reached out presently to Lanty's curls when he entered a little later.

Vashti, white and stately, entered with Sally and took her place in the conspicuous pew set aside for the preacher's family. Sally behaved herself demurely enough in church now, but such is the force of habit that the eyes of all the juveniles in Dole were bent steadily upon the preacher's pew, for in Sally their childish instinct and experience told them there were possibilities, and indeed, to be strictly truthful, it must be confessed that now and then, at decent intervals, Sally treated them to a surreptitious grimace worth watching for.

Mrs. Ranger sat in the body of the church, with the expression of one who perceives an evil odour. This expression was assumed with her Sunday bonnet and laid aside with it. Indeed, Mrs. Ranger thought too much both of her Sunday bonnet and her religious principles to use either of them on week days.

Temperance and Nathan sat alone in a pew well back. It was reported in Dole that they had been seen to look at each other in church, but that was doubtless one of Mrs. Ranger's slanders. Temperance would have been the last to do anything scandalous.

The whole congregation waited.

Sidney was finding his places in the books. This was always an irritating spectacle in Dole, but was forgiven like Sidney's other delinquencies. Dole liked to see the preacher open his Bible with the abrupt air of one seeking a sign from whence to draw his inspiration for the forthcoming sermon. The Dole children had been used to have animated arguments as to whether old Mr. Didymus knew where he was about to open the book or whether his text came to him in the nature of a surprise. If so, then they marvelled that he should so readily find "the bit." Young Tom Shinar had once declared that Mr. Didymus found the place beforehand and substantiated his evidence by saying he had seen little ends of white paper sticking out of the big Bible on the pulpit. But this was coming it too strong for even the most hardened of his adherents, and until Tom rehabilitated himself by thrashing a Brixton boy who said the Brixton church was bigger than the Dole tabernacle, he ran a great risk of finding himself isolated, as sacrilegious people have often been before his time.

To see their preacher searching for his places before their eyes was a most trying spectacle, and no preacher save one of extraordinary confidence in himself and his vocation would have risked bringing himself thus near the level of mortal man. Sidney surmounted this danger nobly, but Dole gave a sigh of relief, as much perhaps for its preacher as itself, when Sidney, after a final flutter of the pages, laid down his books, and rising looked down lovingly upon them; and just as this crisis was reached the door moved a little, wavered on its double hinges, closed, opened again, and finally admitted Ann Serrup, holding her baby in her arms and cowering behind his little form as though it were protection instead of a disgrace. Poor Ann! her *bravado* vanished at the critical moment and left her dazed, frightened, shamed, given into the hands of her enemies, or so it seemed to her. Now the curiosity of Dole over Ann's appearance had been such that there was not one single seat, so far, at least, as she could see, but what held someone. And to advance under the fire of those curious eyes into any of these seats uninvited was more than Ann dare do. Sidney, with the lack of affectation which characterized him, looked about to see the cause of the concentrated gaze of his congregation, and saw a slim, frightened-looking woman standing just within the church door, holding a baby to her breast so tightly that the bewildered child was beginning to rebel against the restraint of the embrace.

Sidney's swift intuitions grasped at once that this was a new comer, a stranger within their gates. He looked towards Vashti—Vashti was looking at the congregation as if expecting one or other of them to do something. Sidney reflected swiftly that it might not be Dole etiquette for the minister's wife to move in such a matter, then he turned to his congregation and said in a voice suggestive of disappointment, "Will not one of you offer a seat to our new sister?"

The effect was electrical.

The Rangers, Smilies, Simpsons, and all their ilk rose at the summons. Ann followed Mr. Simpson up the aisle, but just as she nearly reached the Simpson pew she gave an imploring look at Mabella. Mabella returned an encouraging smile, and Ann darted to Mabella's pew like a rabbit flying to cover. Mr. Simpson felt the defection and resumed his seat feeling he had been "done," and inclined to think Lanty and Mabella had usurped the privileges of the deaconship.

Nathan and Temperance gave a sigh of relief. The moment Ann entered the church each had longed to bring the forlorn girl to their seat, but a kind of shyness had fallen upon these two elderly lovers since their marriage;

retracing the steps of their love dream, they were overtaken now and then by the awkward hesitancies of youth.

Ann put the neatly dressed child down on the seat by Dorothy, and the two babies eyed each other in the frankly questioning manner of innocence.

The congregation recovered, at least outwardly, its equanimity, and Sidney's clear, sweet voice said, "Let us pray," and after an instant's pause uttered a brief invocation to the spirit of Truth and Holiness to descend upon their waiting hearts.

The hymn was sung, and then having read the chapter Sidney closed the Bible and began to speak.

Afterwards when all Sidney's sermons were passed in review, it was remembered that during this discourse he kept his eyes fixed upon the face of his wife, and never once bent his gaze upon his congregation, the congregation which, gathered there full of trust that their spiritual wounds would be bound up, suddenly awakened to the fact that their beloved preacher was smiting them with the cold steel of spiritual condemnation.

This man who had been so ready to empty the vials of healing love upon their bruises, this man in whose hand the spiritual olive branch had blossomed like Aaron's rod that budded, this man whose gentle human sympathy had wiled forth the secrets of the most obdurate, this man had turned and was rending them.

He took for his text, "The sins of the fathers shall be visited upon the children," and ere long the faces before him were piteous.

Never had Sidney spoken as he spoke that day; spurred on, it would seem, by an irresistible inspiration he cried to them "Woe—woe." His fiery words seared their hearts as flame scorches flesh, beneath the burden of his bitter eloquence their spirits fainted. Nor was he content with generalizations, for with striking parallel and unmistakable comparison he illustrated his meaning with incidents from their own lives. Dole had never known how completely their preacher had been in their confidence till he turned traitor and dragged forth the skeletons of their griefs to point the moral of his denunciations.

Beneath it all they sat silent as those mute before a terrible judge, only the swift and piteous changes of expression showed when his barbed shafts struck home. These old men and women were suddenly smitten with the thought that their children who had "gone wrong" were only scapegoats for their parents' sins, sent by them into the wilderness, with a mocking garland of religious training, to take away *their* reproach before the eyes of the world. But though the scapegoat might delude the attention of the world, it did not divert the gaze of the Almighty from their sin-stained souls. Impeached by their preacher's almost personal denunciations, these poor worn old men and women found themselves convicted of, and responsible for, their own and their children's sins. The ghosts of all the bye-gone scandals in Dole rose from the shades, and for once, parading boldly before the face of all men, fastened upon their victims.

Sudden deaths were pronounced to be judgments upon hidden sin, and Mary Shinar's blanched face was wrung when she recalled her saintly father, who was found dead in his field with the whetstone in his hand to sharpen his scythe, that other reaper, whose sickle is always keen, had cut him down without warning. Mary tried to remember what old Mr. Didymus had said about the Lord coming quickly to those whom He loveth, but she could not, she could only writhe under the shadow of a dreadful uncertainty. Death-bed repentances were mocked at as unworthy and unacceptable cowardice, and old Henry Smilie's jaw set, for his son, dead these fifteen years now, had acknowledged between the spasms of the death agony that he had erred and gone astray, and when a merciful interlude of peace was granted him before the end, he spent his last breaths whispering forth prayers to the Saviour whom his life had denied, and when he slept with a kind of unearthly peace and light upon his worn young face, old Mr. Didymus had spoken of those who through many deep waters at last win safe haven, and his father, ground to the earth by heart-breaking toil, wearied by the reproachful tongue of a scolding wife, looked beyond the horizon of this life to that moment when, transfigured from out the semblance of his sins, he should see his only son again. And now—— But Sidney having planted the empoisoned spear in his weary old heart, had turned to other things, and was speaking with strange white-faced fervour of the future.

The congregation had up to this instant rested in spellbound silence, but, as leaving the dead past he entered the hopeful realms of the future and proceeded to lay them waste with the most merciless forebodings, a long suspiration, half moan, half sigh swept about the church, spending itself like a hiss of shame in the corners, and coming vaguely to Sidney's ears, unnoticed at the moment, but to be remembered afterwards in agony of spirit. He made no pause, but continuing in the tense tone of a man who only veils his meaning because it must be veiled, and *wills* that his words be understood, he pictured forth all the terrors which awaited the child of the shamed mother and the child of the drunkard; with pitiless imagery he suggested the inevitability of the fate which awaited them. He denounced in bitter terms the sin of giving children such a heritage, and following out his argument with rigid Calvinistic logic he left little hope of good for the victims of this inheritance. Of all the portions of this bitter sermon, this was the most scathing, and a silence like the silence of the grave fell upon his hearers.

The faces of Temperance and Nathan were wrung with generous, impersonal pain, and they held each other's hands fast clasped, fearing for Mabella, who, her face working with keen, mother anguish, looked at the stony face of her torturer as a lamb might regard the knife which slays it; Ann Serrup, dazed, half stupefied by the storm which beat upon her, had only sufficient intelligence left to shrink from the wounds which followed thick and fast, as a person freezing to death may yet feel the icy rain dashing in his face.

Lanty sat at the end of the pew, a terrible expression of self-reproach in his eyes, his head held erect, his shoulders squared as one who receives the righteous recompense of his sins. But quickening all this endurance into agony was the thought of Mabella, he knew so well what she was suffering.

And, lifted up trustingly, in the midst of these pain-drawn faces, like flowers looking up from amid stones, were the faces of the two children, Dorothy and little Reub.

Having finished their scrutiny of each other they had joined hands and sat silent, looking up wonderingly at the preacher.

Upon their faces there was still the courage and hardihood seen upon the faces of all infants; alas! it is not long before it fades away, abashed by the unconscious recognition of life's terrors. To those who see it, this bravery, the bravery of supreme ignorance, is poignantly touching. And of all that congregation only these two children dared look the preacher confidently in the face.

And yet there was one other. Vashti Lansing, sitting in the extreme corner of the pew, and facing her husband, had never taken her eyes from his face, nor withdrawn her gaze from his.

Her face was white like his, drawn as if by the intensity of concentrated thought.

Seemingly unconscious of the troubled faces about her, yet seeing every variation in their agony, she listened to her own thoughts voiced by Sidney's tongue, she heard her own bitterness translated into words of fatal eloquence.

By the force of her suggestion these ideas, these images, had been impressed upon the mind of her husband, and he read the symbols aloud to his terrified congregation mechanically, only swayed by the more or less emphatic manner in which the thoughts had been suggested to him. And sitting thus, Vashti Lansing saw her own soul face to face.

Surely there must have been something in its dark reflection to terrify this daring woman, surely her heart must have trembled before the magnitude of her triumph, before the spectacle of the misery she had wrought, but if she indeed felt these things she gave no sign. Indeed it would seem as if this woman had suffered so much in secret, over her baulked desire, that she had gone mad of misery, and as some serpents when wounded strike savagely at stones and trees and even at their own coils, so Vashti, in her hour of power, did not care whom she wounded, if so that she could vent her venom and see upon the faces of others some reflex of the agony which had so long lain at her heart.

We cannot explain these things, nor dare we judge of them, for to take judgment upon us is to be ourselves condemned.

Sidney's voice was growing weaker, and finally, with a last scathing rebuke, which was perhaps more of a sneer than a reproof, he sat down, his stern, white face sinking out of sight behind the high pulpit desk.

After a few moments, which seemed a century to the racked congregation, he rose, but the face which they saw was no longer the stern face of the relentless man who had so tortured them. The gentle grey eyes had regained their kindliness, the sensitive mouth its sweetness, the lofty brow was no longer black with condemnation, but bright with beneficence; no longer stern with portents of wrath, but grave with reverent responsibility. He gave out the hymn in his usual way and it was sung haltingly, and then with outstretched hands he blessed his people.

But they wanted none of his blessings. They had trusted him and he had betrayed them into the clutch of their own fears.

It was the custom in Dole for the congregation and preacher to rest a moment or two in silence after the final Amen of the benediction, and after that there were greetings at the church door; but to-day, whilst Sidney's bowed head rested upon his hands he heard hurrying feet crowding to the door, and when he raised his head and descended the short pulpit stairs, he found the church empty, he looked about in amazement.

"Why, Vashti," he said in surprise, "where have they all gone?"

"I don't know whatever has possessed them," she said, although she knew only too well. "But they all hurried out pell mell."

"How strange," said Sidney wonderingly. Vashti looked at him curiously; by this time they were on the porch. It was empty. Those who walked to church had taken their departure, fleeing as from a place accursed. Those who had to wait for the men to bring round the democrat waggons in which they had come from a distance, accompanied the men round to the sheds, and mounting into the vehicles there, drove off rapidly.

Ann Serrup had waited barely till she got to the church door, and then turning with blazing eyes to Mabella, she demanded how she had dared bring her there to be mocked at; the poor tow of Ann's passion was fairly ablaze, but something in Mabella's face quieted her, and with an evil word for the preacher, flung out recklessly from the reservoir of sinful knowledge—Ann departed.

Amid the brief babel of condemnation which had preceded the general departure, the voice of Temperance was the only one raised to stem the flood of popular indignation.

"Perhaps 'twas laid onto him to speak so," said Temperance. "I have heard tell of these things."

"Well," said Mr. Simpson indignantly, "them things is more enjoyable by hearsay. 'Twas disgraceful! that's what it was——" and then he made off, but Temperance, staunch old Temperance, stood her ground, and spoke to Vashti and Sidney as they emerged. But Sidney was wearied out and bewildered by the sudden defection of his people, and so had little to say, and when they reached the little gate the two couples separated and took different roads; the windows and doors were closed in all the houses which Vashti and Sidney passed as they went to the parsonage. Vashti realized that never had she been so identified with her husband as she was that day by the eyes which peeped out of the re-opened doors behind them.

Dole had withdrawn itself from its preacher. It had been hard to win out, but it retired to its shell with a promptitude which suggested that it had never been quite comfortable out of it.

"I can't understand it," said Sidney. "It seems extraordinary. I did not preach too long, did I?"

"No, indeed," said Vashti; "you spoke splendidly."

His face glowed like that of a child which has been praised; he passed his hand vaguely across his brow.

"I am so glad you are pleased," he said. "It was your sermon, you know. It seemed to me I was saying just what you would wish."

"Yes, of course you did," said Vashti as they entered the parsonage gate, then, hesitatingly, she said:

"Have you got it written down?"

"No, oh no—I—the fact is I don't seem to remember what I preached about. How strange! But no matter if you were pleased at the time. I would not care to submit my theology to your tests, my dear."

They were by this time standing together in the little study.

Moved by a sudden tenderness Vashti laid her face against his sleeve.

"I think," she said, "you are better than anyone."

A great joy illumined his face, he put his arms about her, for a moment his old self reasserted itself.

"My dear," he said, "are you well? Why, Vashti, how thin you have grown."

She looked up at him with great hollow grey eyes.

"Thin!" she said, and laughed discordantly; "what should the preacher's wife have to make her thin?"

"You are well and happy?" he asked.

"Both, am I not first lady in Dole?"

"You are First and Only Queen of my heart," he said tenderly. "That's your name and title."

And just then Sally came to say the table was ready, and slipping away from his encircling arms Vashti led the way to the table.

As the afternoon waned, Sidney's nervousness increased. He strove to remember his sermon, and wandered restlessly about the house. At length he came to Vashti where she sat, book in hand, but busy with her own thoughts.

"I'm really worried over the people leaving so to-day," he said. "Can it be that they are disappointed in me?"

"Why, no," said Vashti, then asking a question which had been on the tip of her tongue all day: "Can't you remember *anything* of your sermon?"

"Not a word," said Sidney, "isn't it strange?"

"Oh, it's just a freak of memory," she said.

"Oh, I don't mind that," said Sidney, "but the worrying part is that I seem to remember that I was harsh, that I said cruel things and used the facts you have told me about their own lives to drive in the nails of a cruel argument. Did I do that? Oh, Vashti, tell me. I spoke, it seemed to me, filled with your spirit, so surely I could not have been brutal to them. It is an evil dream."

His pale face was strained with the pain of his thoughts. Vashti was alarmed by the distress upon his countenance. She rose and took him by the hand.

"Lie down, Sidney," she said, "and have a little rest. You are troubling yourself needlessly. Dole is full of freakish people. Temperance has quarrelled with all the women about something and they may have rushed off to avoid some dispute. Your flock think you are perfect. Sleep, Sidney, and forget these troubles. You are too sensitive." He suffered himself to be led to the green leather couch and stretched himself upon it wearily. She bent above him, passing her strong magnetic hands across his brow, looking at him with almost pitiful eyes. Her pity was that of a vivisector who dares sympathize with the dumb creature he tortures.

Sidney looked up at her between the passes of her waving hands. For an instant his face was glorified, and he saw her again as he had seen her that first day on the old porch of the Lansing house, with her fingers shining like ivory in the sun, and her noble head set like a cameo upon the green background of wild cucumber vine which draped the porch.

He saw her thus, his first, last love, and then closed his eyes and floated forth upon the cloud of golden memory into the dreamless realm of a hypnotic sleep with her voice whispering, it seemed within his very soul, saying "Sleep and forget, sleep and forget."

And he slept.

It was dusk. Vashti Lansing let herself out of the parsonage, for a wild hour was coming upon her, the proud, impatient despairing spirit was clamouring at her lips for utterance, and she felt as perhaps every married woman feels sometimes that her home afforded her no sanctuary safe from her husband's intrusion.

And so softly closing the door she fled out into the night, and as her agitation increased, the moonless night deepened, and lighted only by a few wan stars she fled along the country road, her turmoil of spirit translated into physical energy. And presently she found herself opposite the gaunt boulders of Mullein meadow. Its hopelessness suited her mood. She entered it, and wandering amid its dreary boulders she crucified herself with memories.

As a stoic who longs to know the extent of his endurance she forced herself to pass where she had trodden through the furnace, but she did not linger, for deny the fact as she might, Vashti Lansing was no longer the superbly strong woman she had been.

As the "elm tree dies in secret from the core," so Vashti Lansing's strength had been sapped unseen.

She turned dizzily away from the circle of boulders and wandered on, away to the other end of Mullein meadow, and there sank down upon a little knoll known far and near as Witches' Hill, for it was here, so tradition said, that the unholy fires had been lit to torture the life out of cross-grained old women, with perhaps no worse tempers than their judges, but a poorer art in concealing them. It was because of these executions that Mullein meadow was cursed with barrenness, so said the old story, but Dole, concerned with the practical things of food and raiment, did not trouble its head about old tales, only the school children kept the story alive, daring each other to cross Mullein meadow at twilight, or to bring back a stone from Witches' Hill, for there was a strange outcropping of stone here different from any in the district.

Vashti sat beneath the wan sky solitary upon one of these stones. She knew well the reputation of the place, but felt a perverse delight in carrying her tortured heart to the spot where the old Vashti had suffered.

Surely her imperious will, her lawless pride, her revengeful spirit, were as stern judges as those who haled her ancestress to her death.

She sighed aloud, and a wind sprang up and caught the breath and wandered with it up and down the dreary field, till all its barrenness seemed to be complaining to the pitiless heavens of the blight laid upon it.

Vashti rose to depart. As she turned away the wind wailed after her and Mullein meadow seemed to cry aloud for its child to be given back to its stony bosom.

Taking no thought that she might be seen, Vashti crossed to the road, and just as she mounted the fence she heard a cry of terror and saw two figures dash away. The shock to her tense nerves was terrible. She sank to the ground and it was some time ere she regained strength to go on, and when she did, skulking cautiously this time in the shadow of the rough stone fence, she encountered no one.

She reached home, stole into the house, and went to Sidney's room, where he was reading calmly and cheerfully.

So the day ended in outward calm at the parsonage. Two days later Vashti smiled palely when Mabella, who was a timorous and superstitious little soul, told her how all Dole was terrified because old Mr. Simpson and young Ab Ranger, going past Mullein meadow, had seen the ghost of a witch descend from Witches' Hill and come straight towards them. They stood their ground till it began to cross the fence, and then they owned frankly they fled, whereupon it vanished into the earth.

It was described as a very tall, black-robed spectre.

Mabella shuddered as she related this story, and her attitude was typical of the attitude of the whole village. This apparition, seen upon the same day that Sidney had preached his terrible sermon, reduced Dole to a state of consternation. What was coming upon them? Mrs. Ranger, whose belief in and reference to "judgments" was very strong, felt an awesome premonition that a general judgment was in close proximity, and prepared herself for it according to her lights by making up with Temperance and giving Ann Serrup a petticoat.

Having thus hedged as best she could, Mrs. Ranger gave herself up to lugubrious anticipation.

CHAPTER XV

On Monday Dole watched the parsonage gate narrowly, but when Sidney at length came forth he found the little street silent, the doorways dumb, the windows as expressionless as the patch upon a beggar's eye. But silence is often eloquent, speech lurks behind closed lips, and the beggar's patch is frequently only a pretence; as Sidney advanced, the children, playing marbles or hop-scotch in the shade of the houses, rose and ran within, the doors were closed by invisible hands as he drew near, upon the window blinds he could see sometimes the silhouette, sometimes the shadow of a peering face.

Dole had its preacher beneath its most censorious microscope—beneath the lens of prejudice virtues are distorted to the semblance of vices, but beneath the lens of personal disapproval faults become so magnified that the virtues dwindle to mere shadows, and finally vanish. Furtive scrutiny is nearly always condemnatory, and is in its very nature a thing abhorrent; to a sensitive spirit it is simply a sentence of death. The chill of it fell upon Sidney's spirit and weighted its wings as with leaden tears. Coming after the curious circumstance of his people's abrupt departure from the church, Sidney could not but connect their present manifestation of coldness with his sermon.

What had he said? he asked himself, with an agonized effort to force his memory to serve him, but like a spoiled, indulgent servant memory had become a saucy menial and refused to do his bidding. It was impossible for him to dream however that it was the substance of his sermon which had offended them; he had never spoken aught to them but words of peace and hope. It was the spirit doubtless to which they objected. Could it be that, detecting the false ring in his faith, they had turned upon him, as one who had led them from out the wholesome wind-swept places of their stern creed, to the perilous shelter of an oasis of false hope, where they would be crushed in the wreck of the palms of peace, whose stems had no stability, but had sprung up mushroom-like out of human love, instead of spiritual faith?

And such was the innate generosity of this man, that in the midst of his own personal pain, he endured a yet more poignant pain when he thought how their fears and their sorrows would rise to slay them, strong as lions refreshed by rest. He had lulled them to sleep for awhile, was it only that they might gather fresh strength?

One would have said that it would have been an easy matter for a priest beset by these thoughts to vindicate himself before his deacons, but Sidney did not want a hearing. If brought before the bar of their stern orthodoxy what reason could he give why sentence should not be pronounced upon him? And their verdict would break Vashti's heart—the heart which he had striven to satisfy with the gift of his own soul.

Things must go on as they were, he could demand no explanation—nor risk precipitating the expression of any of his deacons' doubts, for he knew, by some blind, unreasoning intuition that his spirit, upon which he had laid such burdens of deceit, would faint utterly before the ordeal. He knew that never again could he force his lips to fashion a false Profession of Faith.

Perhaps his search for the Holy Grail had been an unconscious one, yet he had drawn very near the chalice. However faint his faith in the divinity of the Cup of Christ might be, he yet felt it was far too holy to be profaned by his lips. He abased himself as one who had partaken unworthily.

There is an old parable anent those who pray at the street corners, and he who does not dare even to lift up his eyes.

Sidney turned away from the mute condemnation of the village to the bosom of the hill, and presently found himself over the crest and in the hillside pasture where Lanty's young horses kicked up their heels and tossed their heads, in the arrogant freedom of two-year-olds.

Sidney paused and held out his hands to them, uttering little peculiar calls, and they came to him, at first fearfully, then more confidently, and at last with the boldness of happy ignorance; they did not know yet that man's hand imposes the bridle and the bit.

Sidney had a great fascination for dumb creatures whose instinct distinguishes the real love from the false so much more surely than does our reason. As Sidney stroked their velvety noses, and talked to them, and let them lip his hand, a singular expression overspread his face. For suddenly there faded from it every mark and line imprinted by experience.

The retrospect and dream of love faded from out his eyes and was replaced by the innocent look of the child who enjoys the present moment and anticipates the future with unshaken confidence, the look of one who has neither desired, nor felt, nor yearned, nor suffered. It was a strange thing—such a transformation as one sees sometimes when Death smooths out the furrows and gives back to the worn body the brow of babyhood—signing it with the solemn signet of eternal peace which never shines save above eyes closed for ever. And when our mortal eyes behold this chrism we tremble and call it unearthly, as indeed it is. And this halo shone upon

Sidney's countenance as he fondled the young horses, and talked to them as to brothers, and presently looking at them he began to question them.

"Why is it," he said, "that you have that look in your great soft eyes? I see it always, always in the eyes of you dumb creatures—a look as if you if your hearts were bursting with the thoughts you cannot speak; as if in proud humility you acknowledge that your faculties were maimed—as if you too could render a reason for all that you do, if only you could make it articulate—as if you plead with us to understand you—as if you prayed piteously against the eternal silence which keeps you down. Ah! Do not look at me like that! I know you feel and suffer and think! Look at me as an equal. Surely when you are alone, quite alone, you look at each other with different eyes? Free eyes bright with the unspeakable boon of equality. May I not see you thus? Some night when the moon is high above the tree tops, when the meadow lies like a bright green lake beneath its beams, when the cat-bird calls from the bushes and there is no one here but you, may I not come to you and see you look at each other, and at me, proudly, as brother looks upon brother? For I am your brother! To breathe is to be the brother of all that lives by breath. And see! how ready I am to acknowledge kinship," and so he babbled on and on, all forgotten but the living creatures before him. At length the glory faded from his face, little by little, as a fabric falls into its old folds, his face resumed its normal expression, he patted the outstretched noses all round.

"What piteous eyes you have, poor fellows!" he said, and left them stretching their glossy necks over the fence to him, and pressing their broad breasts against it, till it creaked and cracked.

Dole maintained its attitude unchanged till Wednesday. Upon that day Sidney, passing from the post office, with some books under his arm, met Mrs. Smilie, who, going over to exchange views with Mrs. Simpson about matters in general, and the preacher and the witch's ghost in particular, had left home very early, intending to return before dark.

There would be no more lonely twilight walks taken in Dole for some time to come. The ghost had been seen by several individuals, all testified to its height, its black robe, its white face. Truth to tell, Vashti, dreading to be questioned about her husband's views, had kept herself close within doors all day long, and had taken her constitutionals in the dusk. Did she intentionally play the part of spectre? Perhaps. Nor indeed is it to be wondered at if she grasped at any distraction from her own thoughts, for Vashti Lansing was beset with terrible fears. Working with material she did not understand she had wrought havoc in her husband's brain. His mind had given evidence during the last day or two, not only that it had partially escaped her control, but his own.

Once or twice she had seen the unearthly glory of confident innocence and supernal peace upon his countenance, once or twice his mind had revolted against the charm of her compelling eyes and waving hands; he had apologized for this, as if his will was gone beyond his own control; once or twice she chanced to look at him and met his eyes, and incontinently he fell into a deep slumber.

Vashti's soul fainted within her. How would it end?

Since the Sunday she had avoided any suggestion of making him sleep.

Alas! she had played with fire too long.

Sidney paused to speak pleasantly with Mrs. Smilie, but that good woman did not wish to compromise herself in the eyes of the neighbours by seeming to "side" with the preacher, before she had any idea as to the probable state of the poll. "It will be the first division in the church since long before Mr. Didymus's day," she soliloquized as she proceeded on her way. "I don't believe there would be any division if Temperance and Nathan and Mabella and Lanty wouldn't act up stubborn—but them Lansings!"

These reflections took her as far as her friend's house. The afternoon wore on and Mrs. Smilie was thinking regretfully that it was time for her to get home, and Mrs. Simpson was persuading her to stay with much sincerity, for her larder was full, and Mrs. Smilie was primed with the latest gossip, when there came the sound of voices to the two ladies, and the next moment Mr. Simpson entered accompanied by Mr. Smilie. This solved the problem, both should stay to supper. Mrs. Simpson bustled about with the satisfaction of the housekeeper who knows she can load her table, and presently they sat down and enjoyed themselves hugely over the cold "spare-ribs" and hot biscuits.

After the table was cleared they sat talking some time.

The hour for "suppering up" the horses came. Mr. Simpson rose and Mr. Smilie said they might as well be going, and went with him to get his horse. As they opened the door a faint, yellow glare met their eyes. It lighted up the moonless sky weirdly, and growing every moment brighter, was at length pierced by a long spear of lurid flame.

"Wimmen!" shouted old Mr. Simpson. "Come on; Lanty Lansing's being burned out!"

The two men and women fled along the quiet road in utter silence. A strange hush seemed to have fallen upon the scene, as if all nature's voices were silent before the omnipotent flames which leaped ever higher and higher, as if threatening even the quiet skies. The men and women felt themselves possessed by that strange, chilling excitement which thrills the bravest hearts when confronted by unfettered flame. In the country fire is absolutely the master when once it gains headway, it roars on till it fails for lack of fuel. As they passed the few houses along the way they paused to cry in short-breathed gasps, "Fire! fire!"

Some of the house doors were open to the night, showing their occupants had gone forth hastily; some opened and let out men and women to join the little party of four. The Rangers passed them on horseback, and, as they came within sight of the house, they saw dark forms already flitting before the fiery background, living silhouettes against the flame. It was the great old-fashioned shed which was burning, but the summer wind was blowing straight for the house, and three minutes after the Simpsons arrived a flicker of flame shot out from the coach-house cornice, caught the gable of the old house, crept up it, and fled along the ridge pole like a venomous fiery serpent. Mabella came rushing up to old Mrs. Simpson.

"Will you take care of Dorothy?" she said; "Lanty isn't here—oh, isn't it terrible?" and then she fled back to show the men where the new harness was in the house, and to try to get her sewing machine and a few other of her housewifely treasures. All the neighbourhood was there working with mad energy. These people might gossip and backbite and perhaps misjudge each other sorely, but no need such as this found deaf ears. They knew what such a catastrophe meant, how vital a thing it was, and wild with the energy which is born of hopeless struggle, they strove to cheat the fire-fiend's greedy maw. Ab Ranger and young Shinar were rolling out the barrels of flour from Mabella's well-stocked storeroom, when, high above the noise of the flames and the excited hum of voices, there came the sound of wildly galloping hoofs. The next instant the roan, with Lanty on her back, took the high fence as though it were in her stride, and Lanty, flinging himself from the saddle, rushed to the burning house. He could see for the moment neither wife nor child, nor did he know if the neighbours had arrived in time. He was distraught with apprehension. His wild ride since he had first seen the glimmer of the fire had seemed to him as hours of agony. He ran hither and thither through the crowd uttering incoherent demands for his wife and child.

Mabella appeared in the doorway. The flames lit up his face, distorted with anxiety and terrible fear. A great throb of relief made his heart leap, and released the sanguine blood which rushed to his head.

Mabella and Dorothy were safe—why was he idle?

He leaped towards the doorway, but Mabella, labouring under a deadly apprehension, a terrible fear, had seen his face and been seized by a panic.

"Lanty! Lanty! Don't go in!" she cried.

"Not go in!" he said, and held on his way.

Then a terrible resolution came to Mabella; she had fought bravely to keep up appearances, to hide her husband's delinquencies, now she must betray them to save him. Was she, for paltry pride, to risk letting him enter the burning house in that condition? A thousand times no! He was too dear to her. She caught hold if young Shinar, the strongest man in Dole.

"Oh, Tom!" she cried, "hold Lanty—don't let him go in. *He is not himself.*"

Her voice, shrill with fear and agony, rose above the duller sounds, and pierced every ear there.

Lanty gave an inarticulate sound of grief and wrath and self-reproach. The next moment he felt Shinar's hand upon his shoulder, heard a persuasive if rough voice in his ear, but what it said he did not know, for a wild, blind rage possessed him, and he flung off the hand with a curse. But Shinar would not let him go.

Lanty struck viciously, and the other man called between his teeth:

"Here, Ab—help me hold him"

Ab Ranger came, but it took another yet to hold Lanty, who, perfectly sober, was at length mastered by sheer weight and held helpless, whilst his neighbours strove to rescue what of his goods they could. And then for a little time hot-headed Lanty, moved beyond himself, raved and cursed, and gave colour to any supposition his neighbours cared to adopt regarding his condition. Mabella approached him fearfully, yet her heart was high with the courage which had enabled her to keep him from harm's way. But Lanty with an oath bade her begone. Horrified, she fled to where Mrs. Simpson held Dorothy, and clasping her child in her arms fell upon her knees, crying, from which position she was raised by Sidney's gentle touch. He was white-faced and terribly excited.

"Have you seen Vashti?" he asked Mabella when he had drawn her to her feet.

"No," began Mabella. "I——"

"Here I am," said Vashti in even tones, from near where they stood. "I have been here some little time, but Mabella has been too busy to see me."

Then she turned away and went over to where the men still held Lanty.

"What's the matter?" she demanded, her great eyes blazing, her face white as death.

"Lanty ain't himself," said Ab Ranger.

"You are crazy!" said Vashti contemptuously.

"Mabella said——" began young Shinar.

"Let go of him," said Vashti almost savagely. "How dare you! Lanty is as sober as I am. The idea of you daring to do this thing! They ought to be ashamed, Lanty!"

The detaining hands fell from him. He gave her one look of passionate gratitude, the one sole recompense Vashti Lansing ever received for the love which had ruined her whole life. The young men slunk away. Lanty felt a terrible reaction sweep across him, and fell atrembling with real physical weakness.

He remembered his repulse of Mabella.

"Vashti," he said, "go and ask Mabella to come to me. I said something ugly to her. I want her to forgive me."

Vashti went with seeming readiness. Lanty rested white and trembling, alone, before the flaming ruin of his home. Presently Vashti came towards him slowly.

He raised his head.

"Where's Mabella?" he asked. "She's all right, isn't she?"

"Yes, but, Lanty, I'm very sorry, she won't come."

"Then she can stay," said Lanty heart-brokenly. "If she has the heart to hold out now she can stay; can I come home with you, Vashti?"

"Yes, of course," said Vashti. "I'll call Sidney, and you go home with him. I'll explain to everybody that you are all right. You had better go and not get them all asking questions."

So she brought Sidney, and the two men went away together. As they turned their backs upon the scene there came a terrible crash. They turned and looked.

The roof-tree of Lanty's home had fallen in. He resumed his way with tears brimming his eyes.

Vashti no sooner saw them depart than she hastened over to the group about Mabella. Temperance was holding her in her arms.

As Vashti approached, the group gave way a little.

Mabella looked up.

"What did Lanty say?" she asked eagerly. "Is he ever going to forgive me?"

Vashti answered softly and with seeming hesitation, "Don't take on too hard, Mabella, but he has gone home with Sidney."

Mabella comprehended the words and sank, a dead weight, in Temperance's arms.

Vashti went about in her quiet way, speaking to the oldest women, explaining, or was it only hinting? to them in confidence, how incensed Lanty was against Mabella, how angry Mabella was because of Lanty's words, how Sidney had taken Lanty home to wrestle with him, and how Mabella and Dorothy were going home with Temperance.

Some of the men said they would stay all night, and watch, and gradually the others departed, but even before they separated that night they had found, by the corner of the barn, the point where the fire had been lighted; kerosene oil had been poured upon broken-up shingles, taken from the bundles laid there ready to reshingle the barn when the work grew slack; more than that, Ab Ranger found a box of parlour matches, a luxury little used in Dole; the box was marked with oily fingers.

Who had done this thing?

Mabella, numb with her despair, was taken home by Nathan and Temperance. The tired men whispered together as they lay upon couches improvised of the saved bedding, and watched the embers glow and flicker up into flame, and die away, and leap up again and again.

Vashti was conducted home by the village people.

They stood at the gate watching her run up the little garden path, and open the door of her home; she waved to them from the threshold, and they knew she was safe from the ghost, and as the groups diminished and separated the units composing them drew closer together, for a great fear had laid hold upon Dole.

At length all found sleep, and some from exhaustion, some from despair, some by reason of great grief slept well, but none of them all rested so quietly as did an inert white-faced figure which lay upon the road to Brixton, opposite Witches' Hill in Mullein meadow. A sorrel horse sniffed at the prostrate shape, and whinnied in the night, but it was not till nearly noon the next day that the dead body of Hemans the machine agent was found. His hands and clothes were covered with kerosene oil, in his pocket was another box of parlour matches.

His neck was broken.

The burning of Lanty's home had been terribly avenged.

Vashti Lansing, actuated by the spirit of unrest which possessed her, had taken her big black shawl about her and fled swiftly through the by-ways to Mullein meadow. She had no fears of the night. Her dark spirit was akin to it. In its mystery she saw a simulacrum of the mysteries of her own soul.

And as she sat upon the stones of Witches' Hill and felt the summer wind raising the heavy locks of hair upon her brow, a sense of peace and rest, fleeting, but inexpressibly precious, came to her. Some strange influence made her turn her head and she saw a tongue of flame shoot up like a flaming dart of defiance hurled from earth to heaven. It was Lanty's home! As the thought formulated itself in her brain she was aware of the soft thud, thud, of galloping hoofs coming towards Mullein meadow.

This was the guilty one fleeing from his work.

To think thus was to act. She fled across Mullein meadow to the Brixton road, climbed the fence and crouched in the shadow. As the horse drew near she recognized it in the starlight; knew its rider, and knew her guess was right. Every one knew Hemans' malignant nature, and his emnity towards Lanty was a matter of common report.

The horse was almost abreast of her. She sprang out of the gloom, threw up her arms, the black shawl waved uncertainly about her, the sorrel reared, the man gave a scream of fear and fell upon the stony road striking upon his head. Vashti gathered her shawl about her and fled towards the light which was broadening and glowing against the dusky sky.

Thus Dole was not kept long in suspense as to who had set fire to Lanty's buildings, but the circumstances of his death were hidden from them, but it intensified the superstitious fear which brooded over the village to an agony to think Hemans had been found with his neck broken, exactly upon the spot where young Ranger and Mr. Simpson had first seen the ghost of the witch.

By the afternoon of the following day, Mabella Lansing and the baby Dorothy were installed in the little two-roomed cottage, which alone, of all the buildings upon Lanty's property, had escaped the fire. She had refused all offers of shelter. She would not even stay with Nathan and Temperance.

"I am Lanty's wife," she said, "and as long as there is a roof belonging to him I will live under it. I made a terrible mistake, but some day he will forgive me."

Within her own heart Mabella, great in her love and trust, thought it would not be long till he came to her; she remembered those silent moments in the past when Lanty had made mute acknowledgment of his fault, and she had bestowed voiceless pardon. Mabella knew when she and Lanty met there would be no need for words, and she felt the moment would be too sacred for any other eyes, be they never so loving, to witness.

The first day passed; she saw Lanty at a distance working in the fields. Friday came but did not bring him, and she grew nervous and frightened; the day passed, and the night, but she was growing more and more nervous; she started awake with terror many times during the night; she fancied she saw a face at the window; she thought she heard footsteps round and round the house.

Saturday brought her many visitors, Vashti among the rest. Vashti talked to her about the finding of Hemans' body, the ghost, and the terror which the village lay under, and then departed.

As Saturday waned down to night a sick nervous fear oppressed Mabella; she lit two lamps and tried to fight off her terrors. The ticking of the clock seemed to grow louder and louder. Dorothy tossed in her sleep. Mabella had kept the child awake to cheer her till the little one was thoroughly over-tired. The tension became almost unbearable. She rose, frightened at the sound of her own footstep and took Lanty's violin from the shelf; she could not play, but she thought it would comfort her to pick at the vibrant strings which were so responsive to Lanty's touch. She seated herself beside the lamp—her back to the front door, and facing the door in the rear. She thought she heard a noise behind her—she turned swiftly to look over her shoulder—she caught the shadow of a face at the front window—her eyes dilated. There came a sound from the rear door, and a breath of air. She forced her eyes to look. A tall figure, wrapped in black and with gleaming eyes, stood between the lintels. The fiddle fell; its strings breaking with a shriek. Mabella gave one scream of terror, "Lansing—Lansing!" and darted toward the cot where the child lay—but ere she reached it the front door came crashing in, as Lanty dashed his shoulders against it, and before Mabella quite lost consciousness she felt his strong arms about her, and knew that nothing could harm her.

With Mabella in his arms Lanty rushed across the little kitchen to the empty portal of the rear door, and looked forth, and in the starlight saw his cousin Vashti, with head down, running like a hunted hare for home.

"I know you!" he cried in a clarion-like voice—and Vashti heard.

Lanty, eager, yet ashamed to seek Mabella's pardon, had held lonely vigil without the little cottage; it was his footstep which had so terrified her. It was the fleeting shadow of his face which she had seen. As she looked around he had withdrawn out of sight, and was crouched beside the window when he heard her cry of "Lansing—Lansing!" Only twice before had she called him thus. Once when she came to his arms in Mullein meadow; once during the terrible day when Dorothy came to them, and when Lanty heard it the third time it was as a chord made up of the greatest joy, the greatest agony of his life; he would have crossed the river of death to answer it.

Mabella opened her eyes beneath his kisses. She looked at him, and put up her hand to stroke his face. He caught it and pressed it against his eyes.

They were wet.

"Don't, my dear," she said. "You break my heart," and then the tears so long repressed gushed from her own eyes—and Lanty and Mabella were each other's again—and for ever. And when they were a little calmer they talked together, and each learned how the other had chosen Vashti as an ambassadress of peace.

"Poor Vashti!" said Mabella, a swift comprehension, denied to the stupidity of man, coming to her woman's heart.

"Poor Vashti!" echoed Lanty contemptuously. "Poor Vashti, indeed! Just wait."

"Oh, Lanty," said Mabella with a sob in her voice, "don't *you* condemn her; that would be too cruel."

Lanty said nothing; he had his own thoughts. But the joy of their reunion dwarfed all other interests and peace rested in their hearts.

And Vashti? She had shown no mercy; she expected none. That Lanty would make her name a hissing in Dole she did not doubt.

But so strange is human nature, that Vashti Lansing, confronted with the prospect of shame and mockery for herself, turned to thoughts of her husband. She dreaded the ordeal of the service of the next day upon him. A vague but omnipresent sense of uneasiness, quite apart from dread for herself, weighed upon her. She took a lamp and went into Sidney's room softly; she bent above him. With the stillness of deep sleep upon him he lay very quiet, the delicacy of his clear-cut countenance enhanced rather than modified by the white pillow, and as he slept he smiled. To natures such as his, which harbour neither dislike, distrust nor condemnation of any living thing, sleep is indeed beneficent.

As Vashti looked, slow tears globed her eyes, but did not fall. They were, in all honesty, tears for her husband, not for herself. She bent nearer him and touched him with her lips—perhaps the only time she had ever done so of her own volition.

"I must see him through to-morrow," she murmured—then turning away she left the room. What did she mean? It is hard to pierce to the core of such a woman's soul; but in her great eyes there was the look of one so weary that the prospect of Eternal Sleep seems sweet.

CHAPTER XVI

The next dawn drew from out the dark bright with the portents of a perfect day.

All the hollow heaven was blue as a turquoise stone.

Vashti faced the sunny hours, which yet loomed so black for her, with that courage and calm which grows out of over-much torture.

Pain became its own anæsthetic in course of time—and this numbness had crept over proud Vashti Lansing. She had made others suffer much, but they all had their compensations. Who can say how much she suffered herself?

As the hour for the service drew near, Sidney became very nervous. Vashti tried vainly to console him, but all her soothing words failed to impress him. It was as if she strove to grave an image upon quicksilver.

At last she said to him gently:

"It will be given thee in that hour what thou shalt say."

His face brightened.

"Of course it will," he said simply. "That has happened to me before."

They left the house together. The sun seemed to be more radiant in its revealings than usual that morning, and as Vashti walked down the path its radiance seemed to linger and dwell about her. "A gold frame about the dearest picture upon earth," said Sidney, his loving eyes alight with the adoration of first love. And as he saw her that morning she was very beautiful.

Passing the common height of women she had grown more statuesque and slender, the lithe plastic grace of her girlhood had fixed into a gracious, womanly dignity. Her great grey eyes were profoundly mysterious. They looked out desolately from her tragic face, as the altar lamps of a desecrated temple might shine upon the waste places.

The contours of her strong, beautiful face were solemn and suave as the curves of the Greek Acanthus leaf. Above all there was expressed in her whole face and in every line of her body an intense energy, both of thought and movement. With her to think was to act; to will was to strive.

When distressed by thought or tormented by the behests of her imperious will, she was wont to translate the mental energy into physical exercise, and walk until the demon was laid by physical weariness.

She wore that Sunday morning one of the gowns Sidney had so lovingly designed for her. It fell in quasi-Greek draperies straight from shoulder to heel. Conventionalized by the hand of a clever modiste, it was yet almost classic in its severity. It suited her well.

They arrived at the church a little late.

The congregation was already assembled—and such a congregation! Never in all the annals of Dole had there been such an one. The village had simply emptied itself into the church.

Lanty and Mabella were there, the light of perfect peace and love upon their brows.

Ann Serrup and her baby sat in Mrs. Ranger's pew. That good woman, trembling before the shadow of the "judgment" she was always prophesying, had secured Ann apparently to offer in evidence of good faith, if need arose.

Nathan and Temperance occupied one end of their accustomed pew, crushed into the corner by the overflowing of the unprecedented assembly. And seated in the middle of the church, well back, but just in a line from the pulpit, sat a stranger.

A man with a strong square head, rugged face, and grizzled hair and beard.

A workman, one could see at a glance, and poor as the people of the congregation, but yet there was a subtle difference.

His face was more sophisticated in suffering than theirs—his poverty more poignant—for he knew which they did not, what poor people miss. He had looked wistfully up the highways he might not tread, they looked only upon the hard road they had travelled.

He peered yearningly into paradises of learning whose gates are closed to the man whose hours are spent in toil; they did not lift their eyes beyond the little circle of their immediate needs. He craved to "reach the law within the law"; they sought their own personal salvation.

And as Sidney rose the eyes of this man dwelt upon him as one might look upon a master who had betrayed him, whom yet he follows afar off.

Sidney rose in his place.

A shaft of golden light wavered about the old-fashioned square panes of the window, and, finding the centre of one, pierced through it, and streamed in lucent radiance straight above Sidney's head.

Some in the congregation thought it was like the flaming sword that drove Adam from Paradise, and the old workman, watching the preacher with an infinitude of yearning in his eyes, gave a deep-chested sigh and thought it pity that nature's golden illumination was just a little higher than Sidney, just a hand-breadth beyond him.

With hands outstretched above them Sidney uttered the usual words of his invocation, and then gave out the hymn. There are unwritten canons which govern the selection of sacred songs, and in Dole the clergyman had been wont to begin the service with words suggestive of humbleness, or pleading, or an acknowledgment of the Deity they were addressing, or at least a filial expression of confidence in a Father's love. But Sidney had chosen another hymn than any of these—one of those yearning sweet songs which here and there redeem the hymn-books—usually chosen at the end of the service; he took it as the key-note seemingly of his sermon:

"Oh love that will not let me go."

The congregation sang it wailingly. The preacher rose again and taking for his text these words "Love, the fulfilling of the law," closed the Bible and resting his folded hands upon it began to speak to them, so winningly, so tenderly, that his words smote the flint of their hearts as Moses' rod did the rock. It is one of the terrible tragedies of our imagination to think that the act which saved the wandering querulous tribes alive, condemned the weary old patriarch to only view the promised land. Our souls rebel against the thought, the dispensation seems too bitter, and it is hard to reconcile ourselves to the idea that Sidney, giving the cup of Living Water to these people should himself die athirst—because he had neglected some outward forms. For surely no one could dream but that Sidney's whole life had been one long act of worship.

The old workman had never known before how beautiful the gospel of good tidings might be made. He felt it necessary to steel himself against its insidious charm. Humanized by Sidney's subtle sympathy, and presented to them as a panacea for all human ills, it was little wonder that the old workman began to realize to the full the hold the Christ-word had upon those who believe—though their hearts be rived and strained by earthly cares, though their souls be carded like wool and woven with worldliness, yet there remain ever the little grains of love—the tiny shining particles of faith.

And, as Sidney quoted gentle passages from Holy Writ, a great hope fell upon the old workman that the man preaching these things really believed them—were it otherwise? He shuddered. The magnitude of the hypocrisy necessary for such a deception appalled this disciple of the barren truth. And his hope that Sidney believed was not based only upon the desire to know his idol worthy at least of respect for honesty, if not for judgment; deep down in the soul of this great-hearted man there lived a great love, a great concern for Sidney. He longed to know that Sidney was happy. There was no need to ask if he had suffered. From his appearance it would seem he had suffered almost to the point of death. It would be some compensation if he had won such consolation as he proffered his people. Now this attitude of the old workman's proves his devotion, for it takes a deep, deep love indeed, to make us willing to forget our personal prejudices. But as Sidney proceeded a sick fear fell upon the grey-haired man. For, if unlettered in the higher sense of the word, he yet brought to bear upon any mental question that intuitional acuteness of perception, which in a worthy way corresponds to the natural craftiness which makes comparatively ignorant men so often successful in business.

Nature's lenient mother-heart tries to protect all her children—these gifts seem to be the birthright of the poor. Alas! instead of being used as a defence they are too often upraised in offensive menace.

Beneath the eloquent imagery, the deep human sympathy, the tender lovingness of Sidney's words, the old workman pierced—and found nothing.

Within the sanctuary of Sidney's soul there was no benignant Christ—only the vague splendour of altruistic ideals.

And yet—he held up before his congregation this mask of formulated faith and tricked them as the priests, hidden in the hollow images, tricked the credulous people thousands of years ago.

The old workman almost groaned aloud.

A man of the most lofty mental integrity, this mummery wrung his heart.

"Oh," he said within himself, "if he would only, only once declare the truth—even now if he would cast away these mummy cloths of deception which swathe his spirit. If he would once, only *once* speak and redeem himself for ever."

He looked at Sidney, an agony of entreaty in his eyes, hoping against hope, he looked upon him steadfastly, and suddenly Sidney's voice faltered, a vague expression dimmed his eyes, he repeated himself, hesitated, then in utter silence his eyes roved over his congregation, here and there, as if seeking something definitely defined; and after an interval which keyed up the already tense regard of his hearers almost unendurably, Sidney found the face he sought, and with the unquestioning, unreasoning gladness of a child, he relinquished his eyes to the piteous entreaty in the workman's.

His congregation, whose prejudices had not withstood his eloquence, stirred and wondered, but Sidney heeded not, for the crisis in his life had come.

Who shall explain these things?

In vain the scientist with scalpel and microscope pries and peers, these subtleties puzzle and delude him. For by some curious telepathy, untranslatable in the symbols of spoken speech, Sidney's mind received the impression of the other man's great grief, whose only hope translated itself into a great cry, "Be true; be true."

And Sidney answered it.

For, fixing the attention of his congregation with a gesture as of one who confesses before his judges, he began to speak. And in words of surpassing and subtle eloquence he laid bare every secret of his soul to them. With eyes exalted and glorified he spoke of his love for Vashti Lansing; he told how she had entreated him, how he had hesitated, "but," he said, "her beauty and her goodness stole my soul and I promised to be Minister of Dole."

A swift intaken breath told how Dole comprehended this—the determination to be the minister's wife was easily comprehensible—but the means appalled these people with their faith in the mystic election of priests.

With searching syllables Sidney brought forth the secrets of his soul, and translated to his hearers the doubts and fears, the hopes and ideals which dwelt with him during the period of his long probation.

With face wrung with reminiscent agony he spoke of the day when, after his Profession of Faith, he was solemnly set apart to the service of the God in whom he did not believe.

In some way he made them comprehend his suffering, and a long-drawn groan went up from the over-wrought people, nearly every one of whom had at one time or another agonized beneath "conviction of sin," to whom these spiritual wrestlings were sacred as the birth-pangs of their mothers. With humbleness of spirit he traced his course among them.

He told them in simple touching words of his love for them, of his hopes for the little village in the valley, of his secret plans for their welfare.

Day by day he traced his path among them till he came to the sermon of the preceding Sunday, and, quite suddenly it all came back to him, all its cruelty, its innuendo, its bitter Mosaic logic, writ as in letters of fire upon his heart.

With an exceeding bitter cry he said, "Ah, brothers! This is the evil thing of my ministry. I forgot that the true physician uses the knife as well as the healing unguent. I shrank from paining you, I so eagerly wanted your love; I so dearly coveted your confidences; I so ceaselessly sought your sympathy that I could not bring myself to say anything to wound you. It seems to me that for hard-wrought hands like yours there must be recompense waiting; for weary feet like yours, which have travelled by such stony ways, I thought there must be pleasant paths, and as we are forbidden to take judgment upon us, so doubtless I sinned in judging you so mercifully, but I am too weak to condemn. But my wife, my beautiful wife, more spiritual than I, did not fall into this error, and took the burden from which I shrank. She chose my text for me last Sunday, and when, after reading it, I found myself without words, dumb for very pity before you, suddenly there entered into me the spirit of Vashti, my wife; I cannot explain this to you, but it is true. It was her holy spirit which spake through my unworthy lips."

A quiver shuddered through the congregation; they remembered the old witch-wife—was burning too bitter a penance for such deeds? Silently, but with terrible unanimity, Vashti Lansing was condemned, but their gaze did not wander for a second from the magnetic eyes of their preacher who, with a few more words of eulogy upon Vashti, which were tragically but unconsciously ironic, continued in an almost apologetic way, "I would be the last to question the inspiration of my last Sunday's sermon to you, but yet," more humbly still, as one who, whilst excusing himself, still persists in error, "but yet I can't help thinking we should not dwell too much upon the inclement side of justice; why grieve over sudden deaths when we have read of those who 'were not, for God took them?' Why scorn death-bed repentances when we remember the thief on the cross? Why scoff and turn away from those who sin; why predict generations of shame for them when it is written, 'Though ye have lain among the pots, yet shall ye be as the wings of a dove covered with silver, and her feathers with yellow gold.'" The imagery of the words he had quoted diverted his thoughts to another channel, the apology died from his voice, to be succeeded by the triumph of the high priest who chants a pæan to his divinity, and he uttered an impassioned plea to the men and women before him to endeavour to bring their lives more into accord with the beauty and sublimity of nature, and just as he was soaring into the rhapsodies of pantheistic adoration, there sounded from the elm trees the clear sweet call of a bird.

Sidney paused and listened. It came again.

And then before the wondering eyes of the startled congregation—Sidney's face was transfigured into a semblance of glorified peace. He stood before them smiling in visible beatitude. The sun ray which had been wavering nearer and nearer to him descended upon his brow like an aureola, Nature's golden crown to the soul which adored her; an instant the congregation saw their preacher thus—for the third time the bird's imperatively sweet cry sounded, and Sidney, turning as one who responds to a personal summons, descended the pulpit stair, and following the bird's voice out into the sunshine of the summer day, and was gladly gathered to its bosom. Henceforth he had no part in human hopes or fears. "Heaven lies about us in our infancy," a heaven of infinite freshness, of illimitable joy, of inexhaustible possibilities and gladness.

Sidney's spirit had burst the bars of the prison house and won back to the places of innocent delight, from which each day bears us further away.

Ere he reached the door the grey-haired workman was at his side; there were tears in his eyes—a holy awe upon his countenance, as of one who had witnessed an apotheosis. He wrung Sidney's hand—and Sidney gazed upon him with infinite impersonal loving-kindness—with such a regard one might dream the Deity regarded his creatures.

The workman strove to speak, but the words died in his throat.

"I am so pleased to see you," said Sidney gently. "You have been long away."

"Yes," said the man, "yes—and I must journey on again."

"Then," said Sidney, "I wish you pleasant ways, calm seas and safe haven."

He clasped both the workman's hands in his.

So they parted for ever. The one to tread the hard road down to "the perishing white bones of a poor man's grave."

The other to stray along the golden vistas of ecstatic dreams—till they merged in the dream of death.

And as the workman turned away the congregation came forth and gathered about Sidney; each one in passing the door had turned to give a look of contempt at Vashti where she sat, still and unmoved in her place, and each marvelled at her quietude, but when all the congregation drew from out the church, and yet Vashti did not come, the mothers in Israel went back and found her still sitting there—for she was paralyzed in every limb, though an alert intelligence shone in her great eyes.

They gathered about her, and she confronted them still and silent as another Sphinx with her secret unrevealed. The curse of perpetual inaction had fallen upon her impetuous will; her superb body was shackled by stronger gyves than human ingenuity could devise.

Ah, Vashti! When only a few hours since you had coolly reckoned with the issues of life and death—saying arrogantly "at such and such a time will I lay down the burden of life—and knock unbidden upon death's portal and present myself an unlooked-for guest before his throne."

The poorest of us is wont to say "life at least is mine." How we delude ourselves! We are but infants, priding ourselves upon holding the ends of the reins, whilst Destiny shapes our course—when we would linger in pleasant places we are hurried forward, and when we would flee we are held in some bleak country barren of delight.

They told Sidney gently of what had befallen his wife—but as That Other said "Who is my mother?" so Sidney said, "Who is my wife?" and let his gaze wander to where, high above the housetops, the swallows soared black against the blue....

Mabella and Temperance waited tenderly upon Vashti. Whatever her sins were they were terribly expiated. Through the interminable days and nights she rested there, a living log, imprisoning a spirit fervid as flame, a will as imperious as ever, an intelligence acutely lucid.

We shrink from reckoning up the sum of this woman's torture, augmented by each loathed kindness to which she must submit.

With extraordinary resolution she feigned herself dumb in their hands, from the beginning she had crucified this one of the few faculties left her—she did not choose to be questioned, she would not complain.

She remembered the dream she had had upon the night of her betrothal, and knew that its curse had come upon her.

Lanty sometimes came to her, when she was alone, and told her that he forgave her—that he was sorry for her; he told her again and again, and hoped she understood—but she made no sign—though this all but slew her spirit.

They contrived a wheeled chair for her, and when the weather was fine took her abroad into the sunshine, and sometimes on a summer Sunday, when Lanty did no work, he and Mabella would take her to Mullein meadow, because it was a place of sweet memories to them.

But one grows heartsick at thought of the refined and exquisite tortures this woman endured. Endured unsubdued—for never by one syllable did Vashti break the silence which she had imposed upon her tormented soul.

Dole "hoped against hope" for the restoration of its beloved preacher, but it never came.

He was vowed to the worship of Nature.

At length another preached in his pulpit, an earnest, commonplace man, wise enough to accept with little question accepted truths, only sensitive enough to feel vaguely that he was an alien to the hearts of his people, but attributing the barrier between them to his great superiority. Dole did not forget its duty to the church, but the congregations there were never so great as those which gathered in the churchyard when Sidney came every now and then to talk to them from beneath the elm trees, telling the wonderful truths about Nature, revealing to them in parable the pathos and possibilities of their own lives, bidding them aspire always, expounding to them the miracles writ in letters of flowers upon the hillside, and spelled in starry symbols against the sky. They brought their children to him even as the women brought their babes to be blessed by the Redeemer, and Sidney taught them with unwearied patience, and in more than one instance sowed seed which brought forth a hundredfold. He no longer took solitary walks, for one or other of the Dole children was sent with him always, a happy reverent attendant, whose only duty consisted in suggesting that the dreamer turn towards home at noon or nightfall.

And so we leave Sidney, rapt in the ecstasy of a happy dream, wherein by clairvoyant vision he saw "good in everything."

Nor need we split theological hairs analyzing his claims to mercy.

A mortal genius has said:

"He prayeth best who loveth best

Both man and beast and bird."

And the Christ forgave a great sinner because she "had loved much."

Upon these pleas Sidney's case must rest, if ever he is called before the Grand Jury.

As to the wreck of his mortal life, we can but remember the words of an Eastern martyr, spoken long, long ago—"It is better to be a crystal and be broken than to be a tile upon the housetop and remain."

Made in the USA
Monee, IL
05 November 2026

67383950R00121